Research Ethics for Students in the Social Sciences

Research Ethics for Students in the Social Sciences

Jaap Bos

Research Ethics for Students in the Social Sciences

 Springer

Jaap Bos
Department of Interdisciplinary Social Science
Utrecht University
Utrecht, The Netherlands

With contributions by
Friso Hoeneveld
Utrecht University
Utrecht, The Netherlands

Naomi van Steenbergen
Utrecht University
Utrecht, The Netherlands

Ruud Abma
Utrecht University
Utrecht, The Netherlands

Toon van Meijl
Radboud University
Nijmegen, The Netherlands

Dorota Lepianka
Utrecht University
Utrecht, The Netherlands

ISBN 978-3-030-48414-9   ISBN 978-3-030-48415-6   (eBook)
https://doi.org/10.1007/978-3-030-48415-6

Acknowledgements

I am grateful to Ruud Abma and Toon van Meijl for providing me with insightful case studies (included in Chaps. 5 and 6, respectively) and to Naomi van Steenbergen, Friso Hoeneveld and Dorota Lepianka for co-authoring three chapters (Chaps. 2, 3 and 10).

Versions of this book and of separate chapters have been the subject of discussion over the past few years. We would like to thank those who took the time to read and criticize parts of this book and, perhaps more importantly, encouraged us to continue. A big thank you to Paul Baar, Nienke Boesveld, Susan Branje, Gerrit-Bartus Dielissen, Vincent Duindam, Arjan Eijkelestam, Catrin Finkenauer, Belinda Hibbel, Marcel Hoogenboom, Mariëtte van den Hoven, Jorg Huijding, Cha-Hsuan Liu, Jeroen Janssen, Willem Koops, Karin van Look, Susan te Pas, Isabella Spaans, Jacqueline Tenkink-de Jong, Bert Theunissen, Maykel Verkuyten, Christiane de Waele and John de Wit.

We are grateful to Springer who took it upon themselves to publish this work in open access format. In particular, we thank Floor Oosting and Myriam Poort for encouraging us to pursue this project. Two reviewers, who provided detailed comments and critiques to the manuscript, are to be thanked for their many valuable suggestions and comments.

For granting permission to quote from their ethics guideline, we thank Bronwyn Blackwood from the University of Virginia and Muriel Kaptein from Erasmus University for permission to use samples from their *Ethical Dilemma Game*.

For permission to use his cartoon, we are grateful to John Clark. Thank you Ype Driessen for allowing to use your photo cartoon. For permission to his graph of the Masicampo and Lalande data, we are indebted to Larry Wasserman.

Finally, there have been numerous students who have read earlier versions of chapters and were kind enough to comment on them. In particular, we would like to thank Stefan Gaillard, Emma Heling, Doortje Mennen, Janique Oudbier, Aldith Pasveer, Aris Spanoudis, Lianne Straver and Tim de Vries for their critique and comments.

A shout out to Carys Sterling, who was kind enough to read the video introductions to the chapters.

Special thanks to David Skogerboe for his efforts editing the full manuscript and ensuring it crossed the finish line in style. His improvement saved us from countless errors and mistakes and made this book a much smoother read.

For all remaining errors, great and small, the blame falls entirely on me (JB).

Contents

Part II   Ethics and Misconduct

4   Plagiarism   55

Part III   Ethics and Trust

Abbreviations

AAA	American Anthropological Association
APA	American Psychological Association
BSCT	Behavioral Science Consultation Team
BIT	Behavioural Insights Team
CI	Competing Interests
COI	Conflicts of Interest
COO	Conflicts of Ownership
COPE	Committee on Publication Ethics
COV	Conflicts of Values
CUDOS	Communalism, Universalism, Disinterestedness, Organized Skepticism, Originality
DSM	Diagnostic and Statistical Manual of Mental Disorders
EDI	Eating Disorder Inventory
fMRI	Functional Magnetic Resonance Imaging
GDPR	General Data Protection Regulation
HARK	Hypothesizing After Results are Known
IF	Incidental Findings
IQ	Intelligence Quotient
IRB	Institutional Review Boards
MREC	Medical Research Ethics Committee
MLA	Modern Language Association
MMR	Measles, Mumps, and Rubella
NPM	New Public Management
NGO	Non-Governmental Organizations
PAR	Participatory Action Research
PI	Principal Investigator
PFF	Plagiarism, Fabrication, and Falsifying
PPPR	Post-publication Peer Review
PRFS	Performance-based Research Funding Systems
QPRs	Questionable Research Practices
RCT	Randomized Controlled Trial

RDMP	Research Data Management Plan
REC	Research Ethics Committee
SERE	Survival, Evasion, Resistance, and Escape
SRH	Sexual and Reproductive Health
TESS	Time-sharing Experiments in the Social Science
WEIRD	Western, Educated, Industrialized, Rich, and coming from Democratic cultures

Chapter 1
Introduction

Ethics must be a condition of the world, like logic

Ludwig Wittgenstein (1969, p. 77)

Contents

Keywords Empowerment · Legislation · Perspectives of ethics · Practice-based and problem-based learning · Protocols · Questionable research practices · Referencing policy · Scientific misconduct

About the Reader Let's start with you. The audience for this work is you, a student in the social sciences. Many of the problems discussed in this book will probably be new to you, perhaps not entirely, but still. Yet, right from the beginning of your studies, you have been confronted with certain demands, regulations, and procedures, all driven by certain ethical considerations that you're supposed to be aware of and adopt. You're supposed to be trustworthy, reliable, honest, impartial, and objective if you want to call yourself a researcher. Ah yes, but how? It seems you've got some catching up to do.

Some, such as Steneck (2006), argue that responsible research conduct requires you to learn and follow established protocols and procedures. Others, such as Sim

---

Electronic Supplementary Material: The online version of this chapter (https://doi.org/10.1007/978-3-030-48415-6_1) contains supplementary material, which is available to authorized users.

J. Bos, *Research Ethics for Students in the Social Sciences*,
https://doi.org/10.1007/978-3-030-48415-6_1

1

et al. (2015), insist that your level of engagement and motivation play a role in how you learn and understand research ethics.

In either case, because the rules and regulations of research ethics may appear 'vague' at best, or feel 'beyond your control', we feel that it is important that you are offered an opportunity to see those rules and practices 'in action.'

This book is designed to empower you, to help you grasp research ethics in the most practical sense. By providing you with concrete examples of cases and dilemmas, and confronting them with real questions, we believe you will become more sensible to these problems and will be able to respond to these issues more readily.

Aims and Purpose of This Book Problems regarding research ethics and integrity began to dominate the agenda of social scientists at the beginning of the twenty-first century. Of course, there have always been ethical considerations, but today, more than ever before, we seem aware of the many pitfalls, obstacles, and dangers attached to our research procedures. There are several reasons why this awareness came about in such a relatively short amount of time.

For one, a number of highly controversial cases of scientific fraud within the social science emerged in the early 2000s (among which Diederik Stapel was probably the most prominent). Many of these were widely reported on and helped raise awareness of the dangers of scientific misconduct.

Additionally, and simultaneously, questions were raised regarding what is often referred to as *Questionable Research Practices* (QRPs), which revealed the social sciences' susceptibility to more subtle forms of data manipulation, affecting the field in an unparalleled fashion.

Finally, legislation in many European and North American countries has changed (and continues to change), putting more emphasis on protecting participants, guarding confidentiality, and demanding stringent data management plans.

In the meantime, several outstanding books have been published on research ethics and integrity (see; Resnick 2005; Israel 2014; Koepsell 2016), however, a textbook specifically designed for students in the social sciences remained elusive. When we took it upon ourselves to fill this gap, we reasoned two stipulations needed to be taken into account.

Firstly, if we wanted students to 'get the message', meaning their perspective should be included as much as possible. Secondly, students should be given ample opportunity to 'experience' ethical issues in science as real-life questions or problems, and not so much as abstract rules or guidelines.

This then defines the two goals of this book:

- Inform students about research ethics and raise their overall interest in it.
- Create opportunities for students to engage with ethical problems and dilemmas, allowing them to define their own position.

Educational Plan In each chapter, we introduce the student to the fundamental dilemmas, problems, and choices that one may encounter when doing research. We will focus as much as possible on research conduct, and not on the underlying philosophies of ethics (except briefly in the introductory chapters), or on the ethics of professional conduct (interaction with clients, organizations, etc.). While these subjects fall outside the scope of this work, we will provide introductions for them in subsequent chapters.

Most importantly, we do not offer a 'how-to-do guide.' Instead, the emphasis is on a combination of practice-based and problem-based learning (as opposed to strictly theory-based learning). Our approach rests on the assumption that the student benefits from concrete examples of problems embedded in location and situation specific contexts. Along with a basic understanding of the most important principles and rules that need to be applied, one can acquire this knowledge.

All chapters are written in accordance with the following three-step educational design:

Step one is to *identify* a particular ethical issue as concisely and clearly as possible. At the beginning of each chapter, short informative sections allow the reader to familiarize themselves with basic concepts, theories, viewpoints, and perspectives.

Step two consists of developing *substantiated approaches* to specific problems or issues. How should one address the issues discussed, resolve the dilemmas involved, or avoid getting caught up in them? Short, concrete cases give direct access to the problems at hand without providing moral judgements.

Step three is the accounting or justifying of moral decisions to others. All chapters contain real-life case studies that can be used in class or in tutorials to discuss and probe the choices and decisions.

Structure of This Book We have divided this textbook into four sections that, more or less, represent the various 'orientations' in research ethics, namely a focus on theory, fraud, trust, and formalities, respectively. The division is as follows:

Part I: Perspectives (Chaps. 2 and 3)
Part II: Ethics and Misconduct (Chaps. 4, 5 and 6)
Part III: Ethics and Trust (Chaps. 7, 8, and 9)
Part IV: Forms, Codes, and Types of Regulations (Chap. 10)

The first section presents a brief overview of what science is and what discussions exist in the field of research ethics. There is a short section on what characterizes science, what its outlining principles are, and on different perspectives of ethics. We will focus on three views, each with different assumptions regarding what constitutes moral behavior. The defining question of this section is: *Which perspectives are relevant for the social sciences researcher?*

The second section discusses classical forms of fraud; *Plagiarism, Fabrication, and Falsifying* (PFF). Related to these issues are concerns about cheating, free riding, paper mills, and other fraudulent practices. The defining question here is: *How do forms of fraud impact the social sciences, and why should it be of concern to us?*

The third section deals more broadly with defined issues of research ethics, such as those relating to trust, concerns over confidentiality, conflicts of interest, and questions concerning science and politics. The defining question in this section is: *How should the social sciences define itself in light of the changing political and social landscape today?*

In the final section, which consists of only one chapter, we present a general review and step-by-step discussion of relevant procedures within university codes of conduct, informed consent forms, and other types of regulations found in the social sciences today. The defining question here: *How to design a proper research application?*

A Note About Shaming Over the course of this book, we will discuss ways to tackle ethical issues, sometimes by example of the individuals who chose the incorrect path. This raises the following question: In a book about ethics, is it appropriate to mention the names of those who've crossed the line, committed fraud, or misbehaved in one way or another? Should they be 'named and shamed,' or would it be better to discuss their cases in a more anonymous manner? This is in itself an ethical problem.

We've adopted a pragmatic approach to this question. In some cases, the individual has come to exemplify the problem, such that it would only create an unnecessary distraction were mention of the persons involved avoided. This is true, for example, of the cases of Diederick Stapel, Brian Wansink, and Cyril Burt, among others, which are discussed at some length here. However, if it was at all possible to protect privacy, then we have done so, believing that this principle should prevail.

A Note About Our Referencing Policy It may strike the reader that the authors in this book refer to themselves as a collective 'we' throughout the entire volume, even though only three chapters are authored by multiple writers and two of the case studies were written by different authors, with the remaining majority authored by a single author.

Apart from the fact that it is much more consistent to refer to a single author-identity throughout, there is another reason to speak in the 'majestic plural': all chapters have been read and critiqued by so many different people, who contributed in so many ways, adding so many valuable insights, that it would be almost presumptuous to consider any one chapter the product of a single mind. For this reason, we gladly revive this respected but somewhat forgotten practice.

When referring to unidentified others, we adopt a different policy. At one point it was common practice to use 'he' throughout and forewarn the reader in a footnote that they should understand this as referring to both male and female persons. Later the formula 'he or she' of even 's/he' was adopted. Today, in accordance with the style and grammar guideline in the APA *Publication Manual*, singular 'they' is used when referring to a generic person whose gender is unknown or irrelevant to the context. We've decided to follow this recommendation, as you have perhaps already noticed when in the sentence above we spoke of 'the reader… they' (see Lee 2019 for a discussion this policy).

Beginning, Not the End This book will provide an introduction into research ethics and integrity, but not much beyond that. This is just a beginning, but with two important considerations in mind. First, one will find that many of the questions we carefully separated in this book are anything but separated in real life, and that trying to answer one question has consequences for many other related parts. Ethical questions in real life are rarely simple.

Secondly, one will find that many of the issues discussed in this book are still being debated, and our views on them continue to develop, in part because science itself is in continuous development. Additionally, the fact that science's place in society is changing, how we perceive of ethical questions changes with it.

So, there remains work to be done even after the reader has finished this book. We understand that this may sound somewhat discouraging, but please remember what poet Wislawa Szymborska (2002) wrote in 'A Word on Statistics':

> *Out of a hundred people*
>
> *those who always know better:*
> *fifty-two*
>
> *Unsure of every step:*
> *nearly all the rest.*

This book is dedicated to 'nearly all the rest', namely all those students out there who struggle to do the right thing. We hope this book will help them know how to get there.

References

Israel, J. (2014). *Research ethics and integrity for social scientists*. London: Sage.

Koepsell, D. (2016). *Scientific integrity and research ethics. An approach from the ethos of science*. New York: Springer.

Lee, S. (2019, October 31). *Welcome, singular 'They'*. APA Style and Grammar Guidelines. Retrieved from: https://apastyle.apa.org/blog/singular-they

Resnick, D. B. (2005). *The ethics of science an introduction*. Suffolk: Routledge.

Sim, K., Sum, M. Y., & Navedo, D. (2015). Use of narratives to enhance learning of research ethics in residents and researchers. *BMC Medical Education, 15*(41). https://doi.org/10.1186/s12909-015-0329-y.

Steneck, N. H. (2006). Fostering integrity in research: Definitions, current knowledge, and future directions. *Science and Engineering Ethics, 12*, 53–74. https://doi.org/10.1007/PL00022268.

Szymborska, W. (2002). *Miracle fair. Selected poems of Wislawa Szymborsk* (p. 99). New York: W. W. Norton & Company.

Wittgenstein, L. (1969). *Notebooks, 1914–1916*. London: Blackwell.

Part I
Perspectives

Chapter 2
Science

Contents

After Reading This Chapter, You Will:

- Develop an awareness of what social science is about
- Better understand the role of students within universities
- Recognize the function of reflexivity in the process of knowledge acquisition
- Acknowledge science's institutional imperatives

This chapter has been co-authored by Friso Hoeneveld.

Electronic Supplementary Material: The online version of this chapter (https://doi.org/10.1007/978-3-030-48415-6_2) contains supplementary material, which is available to authorized users.

Keywords Accountability · Bloom's taxonomy of knowledge-based learning
· Communism · Disinterestedness · GDPR · Institutionalization · Learning
community/community of learners · Organized skepticism · Peer review · Perverse
incentive · Positivism · Reflexivity · Universalism · History of science · History of
social science

2.1 Introduction

2.1.1 Our Moral Duty

If you're reading this book, its likely you're a social sciences student. Perhaps
you've only recently embarked upon your journey through the land of the learned
and the learners, or perhaps you're well into your undergraduate education, with
graduate school or the job market waiting just around the corner. All the same, we
hope you're just as excited about becoming a scientist as we were when we embarked
upon our own scientific careers.

Your immersion in science is, as you surely know, part of a larger, collective
human endeavor – understanding and explaining the world in a *scientific way*. As
such, you must approach your work academically, without prejudice or bias and as
free from preconceived ideas as possible. The problem being – this is not
self-evident.

Science in general is about great ideas and technical innovations, but it comes
with a moral duty; to be thoughtful and critical of your own and other people's
ideas. The motto of the British Royal Society, founded in 1660, captures this concept well: *Nullis in verba* (take nobody's word for it).

It is *Nullis in verba,* the skeptical and self-critical approach of the scientific community, that we turn in this book. It is what we, the authors of this volume, but also
the academic community at large, consider the *moral* duty of any scientist (Fig. 2.1).

2.1.2 Understanding of Ethics

In the chapters that follow, we offer an introduction into the ethics of social science
research as an instrument to systematically explore this moral duty of skepticism
and self-critique. We probe the most common moral dilemmas that social scientists
encounter while conducting research, and we discuss several possible solutions to
them, although often no one solution satisfies completely.

Many of the dilemmas discussed in this book are not specific for the social sciences and the questions they raise are common across many disciplines. Different
disciplines struggle with questions regarding how to treat participants in research
with respect, how to ensure that data is collected and stored safely, or how to deal

Fig. 2.1 Motto of British
Royal Society. Source:
Wikicommons

with deception. However, the way these questions are understood and explored may
be different from one field of expertise to the next.

To mention just one example in somewhat more detail: *intrusive research* is a
concern for many scientists. But what is considered 'intrusive' in the social sciences
(something that arouses in the participant unpleasant, even painful experiences or
memories) does not compare with its meaning in the medical sciences (something
that jeopardizes the integrity of the participant's body).

In short, to properly understand the ethical questions of a particular field, we
need to have a grasp of certain 'qualities' within that field. By qualities, we mean a
(historic) understanding of what science and scientific knowledge means to them,
what the aims of their scientific research is, what rules they follow, and if there are
particular questions, discussions, and issues that they are particularly sensitive to.

The purpose of the present and subsequent chapter is to explore the first part of
these 'qualities', while the particular 'sensitivities' will be the subject of discussion
in later chapters.

In this chapter, we briefly explore both the history of science, with a particular
focus on the social sciences, and examine differing perspectives on knowledge. In
the next chapter, we explore a number of important perspectives on ethics, and we
outline several important principles thereof, including a discussion of modern
'codes of conduct'. What these chapters do not offer is an extensive introduction
into the history and philosophy of social science, nor do they extensively discuss
ethics from a philosophical point of view. For a more exhaustive exploration of
these topics, we gladly refer the reader to the ever-expanding lexicon of fascinating
literature on these subjects (see Suggested Reading).

2.2   Science

2.2.1   The Beginning

Europe's first universities date back to the twelfth century, but they were not the first to be founded. Already in the fifth century, ancient universities flourished in India. In Nalanda, for example, the ruins of one of the first great universities in recorded history can be found. It once attracted thousands of students and is believed to have housed a library with over nine million books.

The function of these early universities was principally scholastic, focused upon the articulation and defense of clerical dogmas. However, in a period now known as the 'Scientific Revolution' (sixteenth and seventeenth centuries), the work of knowledge-producing university scholars changed dramatically in Western Europe (Fig. 2.2).

It was as if the human imagination had been liberated. From this period onward, natural philosophers (the precursor to 'scientists', a term first used in the nineteenth century) were allowed to 'wonder' without dogma – performing experiments conceived first in the mind and controlled through rationality. The mystery of the cosmos offered more than a feeling of awe and amazement, becoming a backdrop for a cascade of questions: Why do the celestial bodies move in the way they do? What makes them move? What are they even? What is light? What are the 'natural forces'?

Fig. 2.2 *University of Nalanda.* (Source: Wikicommons)

How large is the universe? How old is it? Where does humanity fit into it all? How do we *think*? Why do we think? What am I? I think therefore I am – right? Minds were being blown, one question at a time.

These broad questions that helped form the basis for what we today call 'science' have kept generations of scholars busy ever since. But what do we mean exactly when we talk about 'science'?

A common characterization of science is that it's an attempt to explain reality and offer knowledge that can help predict or prepare for future events. But is this unique to science? Note that there are other organized systems of though (such a world religions) that aim to do the same.

A more elaborate answer would be that science (a) produces a body of robust knowledge by way of (b) a certain methodology, and it does so within (c) an infrastructure of physical institutions (such as universities, laboratories, etc.).

These three dimensions of science (knowledge, methods, and infrastructure) presuppose a fourth dimension, which is particularly relevant in the context of this book. Knowledge, methods, and infrastructure require (d) a set of *moral values*, embedded in our academic way of thinking. Moral values structure the scientists' activities. Producing robust knowledge within the framework of an institution means you must adhere to certain rules, regulations, and appropriate methodologies.

On one hand, the procedures governing the act of actually 'doing' science are institutionalized in the regulations and protocols of each discipline – a moral compass defined on paper. On the other hand, they are more implicit, with greater reliance on the moral virtues of the individual researcher and are thus more difficult to identify (more on this in Chap. 3). Both ways of considering the set of moral values in science, institutionalized and implicit, are played out on the center stage of this book, representing the ethics of scientific research and the integrity of the researcher, respectively.

In the next section, we briefly outline the above-mentioned dimensions of science – knowledge, methodology and infrastructure – as framed against the background of the developing social sciences as they emerged in the nineteenth and twentieth centuries.

2.2.2 A Very Brief History of the Social Sciences

Before the social sciences entered the academic arena in the nineteenth century, the 'project of science' was closely connected to the natural sciences. From Newton's law of gravity to x-rays, and from gun powder to penicillin, science was all about great discoveries. Even the idea of 'discovery' is connected to science: the very concept did not exist before the Scientific Revolution (Wootton 2015).

Powered by science and its instruments, such as compasses, canons, and cartography, Western countries set their sights on world domination and established, largely during the long nineteenth century of colonialism, their global empires. The scientists themselves were predominantly white males from a privileged background.

From the mid-nineteenth century onward, several disciplines professionalized and institutionalized into different *subdisciplines*. For each separate field of scientific inquiry, particular methodologies were prescribed, and dedicated societies and journals were founded. Astronomers built telescopes and observatories to study the stars and began to develop theories on the origin of life; biologists probed the world unseen by human eyes with microscopes and developed specializations such as botany and zoology. Nearly every other knowledge producing discipline followed similar patterns and scientific inquiry developed in divergent directions.

Within this large network, most of the actors shared the *positivistic* ideal, meaning they believed that progression in science is understood as an accumulation of true and empirically confirmed, factual knowledge. Science had become not only a new arbiter in matters of truth and falsehood, but it was also seen as a strong instrument for improving the human condition. It had taken up a position which was previously, and rather exclusively, the realm of religious systems.

It was during the nineteenth century that the *social sciences* stepped foot on the stage. The social sciences first emerged in the shape of political economy, sociology, and what was then called the 'moral sciences' (an early form of psychology). Early social scientists sought to transform the rising nation-states of the world into stable, governable economies.

The second half of the nineteenth century revealed a need for analytical insights into the inner-workings of capitalism, the state, and its growing bureaucracies. The quest for this knowledge laid the foundation for the modern-day social sciences. In order to quantify human behavior and to get a grip on the emerging patterns in modern societies, they employed their own discipline specific tools, such as statistics, which proved to be a valuable instrument for their cause.

By the early twentieth century, social scientists were already studying a multitude of topics, spanning a wide-breadth of human-related matters; from perception and consciousness, psychopathology, and public administration, to problems of recruitment and selection, the mysteries of religion, and the supernatural. From these different areas of interest, a variety of new disciplines, subdisciplines, and schools of thought emerged.

For example, within psychology in the 1920s alone, there were Gestalt psychologists, behaviorists, experimental psychologists, industrial psychologists, even 'parapsychologists' (who studied the spiritual dimension of life), not to mention psychoanalysts (who had their roots in medicine). All these subdisciplines and their corresponding schools of thought developed their own institutions, established their own journals, and formed their own methodologies.

Similar developments took place in sociology, anthropology, and economics, as well as in philosophy, history, and theology, all disciplines that were then still considered bastions of the social sciences. A number of subdisciplines that formed during this time, such as what we would now call clinical psychology and neuropsychology, were not yet regarded as a part of the social sciences, but rather part of psychiatry. Educational studies were only in their infancy, and political science, gender studies, and interdisciplinary studies would not emerge until much later, generally after World War Two (Repko et al. 2014). Which disciplines belonged to the social sciences and which did not has long been a subject of debate, raging still today. This illustrates the fact that the social sciences as a whole are still a collection of rather loosely connected fields of developing knowledge.

By the second half of the twentieth century, two developments further shaped the field of social science. For one, the social sciences had become regarded as an independent 'discipline' and was no longer considered the offspring of other disciplines (almost all early psychologists trained in the nineteenth century were physicians, for example). While the methods and corresponding 'objects of knowledge' the social sciences sought were situated between the natural sciences (explaining the world by means of natural laws and experimentation, resulting in objective knowledge) and the humanities (understanding the world with ideographic methods, resulting in more subjective narratives), their object, human behavior, was unique.

Secondly, a strong impetus towards independence came via a post-war surge of popularity in the social sciences. There had been only a handful of students interested in psychology or sociology in the years prior to 1940, but this dramatically changed in the 1950s, ramping up further from the 1960s on. Thousands of students began enrolling in social science disciplines like psychology, sociology, educational sciences, and political science to meet the growing demand for social scientists. Applied science became one of the social sciences' most valuable additions, delivering an innumerable number of new therapists, educationalists, human recourse managers, test psychologists, and policy makers every year.

This rapid influx allowed the social sciences to establish itself firmly in the post-war framework of modern universities, which persists today. Scores of professorships were created, large research institutions were established, and considerable sums of money began flowing into the social sciences. These processes of institutionalization and professionalization went hand in hand with the formalization of research procedures, reflected in stricter and more formalized views on ethics, exemplifying an increased concern with scientific misconduct (discussed in Chaps. 4, 5 and 6).

Approaching the end of the twentieth century and beginning of the twenty-first, new developments set in motion a series of changes that transformed the outlines of the social sciences once again. Neoliberal politics caused budgets to dwindle, forcing social scientists to collaborate with other disciplines, cross-pollinating their work. Many strove to 'valorize' their work, emphasizing its commercial value and thus allowing others to influence their research agendas, intentionally or not.

These tendencies, though grossly oversimplified here, clearly reflected on the social sciences' fundamental commitment to understanding the world. While some argued that the social sciences were acquiring a newfound importance in society, others doubted that the knowledge it produced was capable of withstanding tests of validity, and in response a 'replication crisis' was declared, a charge many sought to counter (Nussbaum 2010). We will return to these arguments in Chaps. 8 and 9 (Box 2.1).

Box 2.1: 'What Do Social Sciences Study?'
The Study of Humans Debates about the validity of social science knowledge exemplify the challenges in the scholastic study of humans. One of the
oldest and arguably most notable disciplines in the social sciences, sociology,
focuses on collective human activity, social relationships, and social interaction. As it is situated at the interplay between social structure and individual
agency, one of its most fundamental issues lies in the existence of social structures and how they objectively influence our lives.

Psychology, on the other hand, often seeks to understand and predict individual human behavior in a way resembling the 'hard' (natural) sciences. The
working of the mind, cognitive processes, and functionality of the brain have
all been the subject of psychological research. Subdisciplines such as neuropsychology, developmental psychology, social psychology, and clinical psychology are all devoted to different dimensions of individual behavior.

Taking a longer view, anthropology studies the rituals, values, and practices of human societies and cultures, forming subdisciplines in cultural,
social, medical, and linguistic anthropology. Because cross-cultural analysis
plays such an important role in the study of anthropology, questions regarding
cultural relativism (to what extent are someone's values to be understood as a
product of their culture) have always received a lot of attention from university scholars.

2.3   Knowledge

2.3.1   The Role of Universities

If science's most important task is the production and reproduction of knowledge by
use of certified methodologies within a structural framework of institutions, then
how should universities prepare students for this feat?

Gabelnick (1990) proposes that we should view universities as *learning communities*. Universities are institutions populated by professors, teachers, researchers, staffers, managers, and of course students, who are all committed to the same
objective – the accurate production and reproduction of knowledge.

The perspective that universities are learning communities takes for granted that
academic institutions are bureaucratic organizations seeking cost-efficiency. In
order to do what they must do, they should have strict curricular structures, consisting of well-defined teaching programs with formal learning objectives, prescribed
assessment criteria, and quality control agencies. Such environments strive for productivity as their goal, allowing for little deviation from the norms they put in place.

Indeed, modern teaching programs at universities often wield knowledge as their instrument and regard students as passive consumers of it.

Dissatisfied with such a restricted view of the student's role in the university, Etienne Wenger proposed an alternative perspective. His work has been influential in higher education circles since the 1990s. Instead of regarding learning as a formalized activity, carried out by isolated members of an institution, Wenger proposed that learning is a *shared* and *situated* activity that requires *communities of practice*.

In communities of practice, people are actively engaged with each other, constructing knowledge together. Participants in these communities 'share a concern, a set of problems, or a passion about a topic, and deepen their knowledge and expertise in this area by interacting on an ongoing basis' (Wenger et al. 2002, p. 4).

Participation, sharing, and interacting in communities of practice are essential elements in learning, since it is *through* participation that identity and practices develop. Participants in communities of practices 'learn by doing' (instead of learning by absorbing or consuming knowledge).

In this book, we too adhere to such a *constructivist perspective,* and we invite the reader to be actively involved with the normative questions raised here, developing their own solutions to moral dilemmas. Of course, books are interactive in only a limited sense, but hopefully the case studies offered in the following chapters, along with the corresponding exercises that accompany the chapters, enable students to become involved in these debates. We want them to be able to discuss their ideas and engage with classmates, co-constructing their own solutions to the problems posed here (Fig. 2.3).

Fig. 2.3 Communities of practice

2.3.2 Knowledge Construction

If your role at university is to 'co-construct knowledge', then what exactly should you develop? What does true knowledge consist of?

Fundamentally, knowledge is simply *any information* about the world (i.e. 'Moscow is the capital of Russia', 'Water consists of two components of hydrogen and one component of oxygen'). In an academic context, however, knowledge is more precisely defined as (a) a body of discipline-based theories, concepts, and methodologies, and (b) any number of practical generalizations and principles that apply to fields of professional action (Eraut 1994, p. 43).

Thus, what psychologists, sociologists, and anthropologists claim to know is a result of how they *define the world* and of how they *operate in their fields of research*. Accordingly, not just their view of the world, but the properties they ascribe *to* the world or *in* the universe may differ radically from one discipline to the next.

Acquiring knowledge implies much more than simply learning a set of theories and concepts. Being immersed in academic education, you'll follow a process of gradual mastery.

At first, learning is about understanding the fundamentals of a field of knowledge. At this stage, little ownership is involved. Research procedures are learned, reporting preferences are practiced, and the existing historiography is read.

Quickly thereafter, these fundamentals need to be applied to practical situations, and the knowledge of other disciplines becomes indispensable. An increased sensitivity to the explicit and implicit norms and expectations across disciplines becomes a tool for collaboration. Despite a sharp learning curve, by the end of your education, you are expected to display analytical skills, propose your own ideas, and develop insights in your own right. Only then have you become a trusted and productive member of the academic community, a co-constructor of knowledge.

Benjamin Bloom's taxonomy of cognitive knowledge-based learning attempts to grasp this gradual development (see Figs. 2.2 and 2.4).

2.3.3 Risk and Reflexivity

A key factor in becoming a 'trusted and productive member of the academic community' is *reflexivity*: the ability to critically reflect on the responsibilities of both yourself and others.

Reflexivity isn't just some invitation to be cautious or thoughtful. We are living in an age of increased *accountability*, meaning that more than ever, there is an obligation on individuals, businesses, and institutions to explain and justify the choices they make. For scientific researchers, this means that you can and will be held accountable, or even be liable, in a case of wrongdoing, intentional or not.

Fig. 2.4 Revised
taxonomy of Bloom's
knowledge-based learning.
(In Anderson et al. 2001)

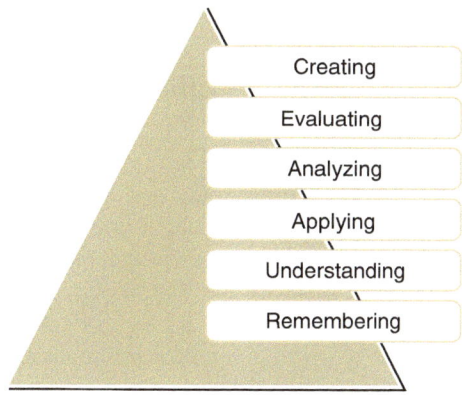

Creating

Evaluating

Analyzing

Applying

Understanding

Remembering

Box 2.2: 'GDPR'

Under the European *General Data Protection Regulation,* individuals have the right to access their personal data, the right to be informed and/or forgotten by those who use their data, the right to object or restrict (further) use of their data, and the right to be notified in case a data breach has taken place.

GDPR requires that concrete and appropriate procedural and technical measures be taken to protect these rights. Enacted in May 2018, it has far-reaching consequences for all institutions (including universities) that use personal data. Institutions are required to:

- Create a comprehensive privacy policy;
- Appoint data protection officers and representatives;
- Adopt specific codes of conduct;
- Maintain records of all data processing activities.

The monitoring of risk is therefore a crucial aspect of reflexivity (Giddens 1991). Risk assessment is no longer an individual responsibility, but a collectively carried burden, and this reality had grown in significance over the years.

Much of this collective responsibility has been written into regulation at an international level. In Europe, for example, the *General Data Protection Regulation* (or GDPR for short; see Box 2.2), constitutes a set of binding directives that protect the rights of human participants in research, ensuring that researchers and research institutions actively assume responsibility.

At the level of local institutions, special independent controlling bodies have been installed to coordinate the ethical dimensions of research. Most universities and research institutions today require that researchers submit their proposals to these *Institutional Review Boards* (IRBs), who demand strict procedures when considering research applications. Funding agencies will often demand compliance with these bodies, and many journals require approval from them before they publish an article.

2.4 Ethos

2.4.1 Science's Ethos

We started this chapter with the observation that science's mission is to understand the world systematically and methodically, while being as unbiased as possible. Furthermore, every scientist has a moral duty to be skeptical and critical. We now return to this grounding principle and ask: Are there any general guidelines that scientists must follow that allows them to be both critical and methodical? Yes, there are.

One of the first demands scientists must meet is the need to remain autonomous. It has been long advocated that universities should safeguard their independence. They must seek *objectivity* and establish as much self-regulation as possible. Science is not to serve interested parties; either influenced by ideologically, politically, or commercially inspired motives.

Another important principal is that scientists must fulfill their tasks carefully and *reliably*. The methods and procedures of science should be transparent, its studies replicable, and its results accessible to all. This open character of scientific knowledge is pivotal to its mission.

In his now famous 1942 article 'A Note on Science and Democracy', American sociologist Robert Merton formulated several essential principles which, if followed, he argued would ensure science a secure and autonomous place in society. He wrote this at a time when Western civilization was at the threshold of being radically transformed politically, culturally, and economically, with the free exercise of science all but self-evident. As such, Merton (1942) proposed four 'imperatives' that make up the 'ethos of science' (Fig. 2.5).

1. *Communism*, later dubbed communalism. Because knowledge is the product of collective effort, substantive findings of science are assigned to the community. They constitute a common heritage and therefore 'property rights' are held down to a bare minimum.
2. *Universalism*. The acceptance or rejection of scientific claims should not depend on any personal or social attributes of the researcher. Only pre-established, impersonal criteria should be used to determine a truth claim.
3. *Disinterestedness*. Scientists should act for the benefit of a common scientific enterprise, not for their own gain. Self-interest should play no part in science.
4. *Organized Skepticism*. Scientists should be skeptics, suspending judgement until the facts are at hand. Logical and empirical criteria allow for a detached inspection of any claims, which are exposed to critical scrutiny before being accepted.
5. Later a fifth imperative was added: *Originality*. Researchers must create new scientific knowledge, and not just reproduce established findings (Ziman 2000).

These five imperatives are known by its acronym CUDOS. These imperatives have found their way into various 'codes of conduct' (to be discussed in the next chapter) and existing scientific practices and procedures (see the last chapter of this book for a detailed discussion).

Fig. 2.5 Robert K. Merton, at Leiden University, the Netherlands, at the occasion of receiving an honorary doctorate, 2 July 1965. Source: Wikicommons

The principle of communism, for example, is accomplished through the practice of academic publishing, which allows researchers to share their findings. The principles of universalism and disinterestedness then, are grounded in the widely accepted practice of 'peer review.'

The all-important practice of peer reviewing (which means that authors submit their work to a forum of experts before it gets published) was not yet utilized during Merton's time, but gradually became standard practice in the decades after World War Two. From then on, it was the reviewers who decided whether or not a paper met accepted standards. Reviews are as a rule 'blind', which means the identity of the author remains unknown to the reviewers, ensuring fair judgement. Peer reviewing itself is considered part of one's 'academic duty,' and the imperative of organized skepticism, for reviewers don't get paid (Box 2.3).

2.4.2 Ethos or Arena?

The question that emerges at the end of this chapter is whether these Mertonian guidelines (communalism, universalism, disinterestedness, organized skepticism, and originality; CUDOS) still meet our expectations. As we indicated in our brief history of the social sciences, much has changed in the past two or three decades. So then, what is the value of scientific knowledge, however 'true', if it isn't supported by politicians, policymakers, or other stakeholders?

Social science in the early twenty-first century feels like a battle in an arena. And this arena is anything but a level playing field. It is populated not only by

Box 2.3: 'Climate Change or Nudging. A Dilemma'
Since the early 2000s, climate change – and the totality of challenges associated with climatic variability and change – is recognized among scientists as an indisputable fact, even though there may be debate among researchers about specific causes, implications, or future scenarios.

In an attempt to anticipate some of the most pressing issues related to climate change, scientists have proposed a wide range of ideas and solutions that could be implemented by policymakers. Social scientists have also contributed their fair share to these solutions. The challenge they face is in finding ways to alter human behavior on a massive scale. In response, the UK-based Behavioral Insights Team (BIT) was founded in 2010. One of the solutions they developed were small, low-cost 'nudges' that focus on making subtle changes to people's environments. For example, loft insulation helps reduce energy waste, but few people were installing it. When provided with low-cost labor to clear their lofts, however, the number of loft insulation installations increased fivefold.

Nudges may only impart a slight change in human behavior, but they are cheap to implement and when used on a large scale, can still have a significant impact. Nudges are contrasted with traditional government levers for behavior change, that include vast mechanisms for interfering in economic and ecological systems, many of which are much more resource and cost intensive.

Recently, Irish Prime Minster Leo Varadkar adopted the 'BIT approach' and argued that the government's pathway to zero emissions by 2050 was to 'nudge people and businesses to change behavior and adapt new technologies through incentives, disincentives, regulations and information' (*The Irish Times*, June 17th 2019).

We may all agree that climate change is real, and that 'something needs to be done' about it, but the question here is: what role should social scientists play in this discussion, and whether or not they should get involved with politics.

Proponents could argue that the social sciences have a moral obligation to invent instruments that help drive human behavior in the right direction. Opponents could argue that nudging does not involve consent – people are 'gently pushed' in a certain direction and may not even be aware of it. They may contend that it's not up to the social sciences to invent instruments that steer the behaviors of people, it should be up to each individual to make their own decisions.

Where do you stand in this debate?

professional and highly competitive scientists, but also by networks of stakeholders, financers, managers, journalists, and career officers, all operating with their own interests in mind.

These developments, while grossly oversimplified here (we offer greater attention to some of them in later chapters), clearly threaten to undermine science's fundamental commitment to an unbiased, systematic, and methodical understanding of the world.

True, science's mission has remained largely the same throughout the past few centuries, but that doesn't mean its mission ever was nor is self-evident today. To accomplish its goals, science must relate to society, adapting to its needs and wants, its political, and even commercial pressures. At the same time, it must seek to retain its integrity and autonomy. And for this, we need ethical reflection.

2.5 Conclusions

2.5.1 Summary

This chapter started out with a definition of science's original mission: to understand the world without prejudice or bias, free from preconceived ideas or dogmas. From this, an accompanying obligation was derived, to be *skeptical* towards one's ideas and the ideas of others.

A brief summary of the history of science and the history of the social sciences outlined a picture of a field still in development. A view of universities as *learning communities* was contrasted with one of universities as *communities of practice.* Furthermore, a *constructivist perspective,* in which students are regarded as active co-constructers of knowledge, was proposed. It was argued that being immersed in academic education implies a process of gradual mastery and increasing ownership, in which *reflexivity* plays a crucial role.

Finally, Merton's 'institutional imperatives' CUDOS were discussed: *universalism, communism, disinterestedness* and *organized skepticism,* later extended with *originality,* and how they extend into scientific practices such as peer review procedures.

2.5.2 Discussion

We investigated the 'parameters' of scientific practice, and we questioned the conditions that enable it. We found that universities as institutions have changed since their early days, though at its core the mission of the scientist has remained the same: to understand and explain the world, free from bias and dogmas. We further identified one value that we considered pivotal for science to fulfil its mission: to remain autonomous, at least to some degree. One question we did not ask: how will science succeed in remaining autonomous? The answer can be found in the subject matter of this book: its ethics.

Case Study: Max Weber and the Calling of Science

In a 1917 lecture before a group of politically left-leaning students, sociologist Max Weber spoke on the particularities of a career in modern science. What should students expect to find in academia and what is expected from them, what should motivate them, what shouldn't, and what responsibilities do they have? (Fig. 2.6).

In this lecture (published in 1919 as 'Science as a Vocation') Weber first used the expression 'disenchantment of the world,' which roughly points to the process of rationalization and simultaneous erosion of the sway religion and superstition held over humanity since the Enlightenment. It is a process, says Weber, to which science (understood in the broad sense, as both the natural sciences, the social sciences, and the humanities) acts as 'a link and a motive force' (Weber 1991, p. 139). Science is thus both a product of rationalization and a driver of it as well.

What is the task of a scientist in the 'modern' (early twentieth century) world? Weber discussed several dimensions. For one, science depends much less on

Fig. 2.6 Photo Max Weber, 1864–1920. (Photo credit: Wikicommons)

creativity, 'enthusiasm' (passion), and inspired ideas than it does on hard work, and above all, specialization. 'A really definitive and good accomplishment [in science] is today always a specialized accomplishment' (p. 135). The difference between a creative outsider and a scientist is not who generates good ideas, it's that the scientist has a 'firm and reliable work procedure' that the creative outsider may not.

Indeed, the positive contribution of science in one's life consisted of: (1) the production of knowledge, (2) by use of reliable instruments (such as the experiment) and sound methodology (logic), (3) to create clarity, (4) in an attempt to serve the truth. Science is a 'vocation', Weber insisted, 'organized in special disciplines in the service of self-clarification and knowledge of interrelated facts. It is not the gift of grace of seers and prophets dispersing sacred values and revelations […]' (p. 152).

Science does not, therefore, dabble in interpretations of 'meanings', neither of a political or religious nature. A scientist is a researcher and a teacher, not a 'spiritual leader', and although they may hold certain convictions, they should not be communicated in class. 'The true teacher will be aware of imposing from the [university] platform any political position upon the student. […]' (p. 146).

Furthermore, Weber spoke of 'disinterested science,' the imperative that scientists should act not in their own self-interests, but for the benefit of a common scientific enterprise, and that the scientist is just a part of a larger quest for knowledge. A quest in which all its products (knowledge) have a short shelf life. Every scientist knows that their work will be outdated in 10 or 20 years. It is the only way to make progress. That is science's fate, or better yet, its very meaning.

Whether science as a 'vocation' is worthwhile under such circumstances is a question that cannot be answered, because it implies a values judgement, and science does not play those games. However, at the onset of his lecture, Weber made it quite clear that the career prospects of a young scholar in the early twentieth century were not very good. Comparing Germany with the US, Weber identified that the career of a German scholar depended on luck and savvy networking, not ability. In the US on the other hand, academic careers were subjected to bureaucratic procedures that created less uncertainty, but also offered less academic freedom. In both cases, young academics had to work long hours, receive meager salaries, and were anything but certain whether they would be promoted. 'Academic life is a mad hazard', Weber mused (p. 134).

Over a century has now passed since Weber's lecture. The academic world, and indeed the world at large, has since changed. Do the conditions for aspiring scientists mentioned by Weber still apply today, and are the values he professed still recognized? Hackett (1990) probed this question in 1990, contending that the circumstances at universities have changed dramatically, and that as a result, several major underlying values have changed too. Let's briefly look into his argument.

Hackett argues that (a) universities have become increasingly dependent on recourses from the private sector, which has caused universities to closer resemble the private sector, and (b) universities have become increasingly dependent on government agencies, causing universities to behave more like bureaucracies. These 'isomorphic forces' (constraining processes that force one organization to resemble another) have shaped and structured university cultures in significant ways, creating ambivalence and value conflicts. Hackett claims that these conflicts may lead to social disorganization and even scientific misconduct (1990, p. 249). Drawing on interviews with scientists, chair holders, academic administrators, and other officials, Hackett points out a series of 'value tensions', of which we will discuss three:

1. *Academic freedom and autonomy versus accountability.* In Weber's time, academic freedom was still more or less guaranteed, but in the 1990s, academic scientists became 'more accountable to and directed by their research sponsors' (p. 266). This reflects in their choices of research problems and publication practices.
2. *Research versus education.* Principles and practices that the scientist as mentor would prefer (to stimulate free and independent inquiry) have become inconsistent with the needs of the scientist as employer (to 'produce' research for a market).
3. *Efficiency versus effectiveness.* Pressure to perform within tight budgets and schedules undermine the scientist's ability to deliver quality work. Measurable standards of performance and detailed accounting strip science of the aura of expertise (p. 269).

Weber predicted that universities would follow the American path, towards bureaucratization, and from Hackett's observations, it seems he was right. 'Universities, scientists […] and graduate students are assembled in a cascade of dependence, which sharply contrasts with the ideal of independence in science and academe', Hackett notes (1991, p. 270). In Chap. 9, we continue this discussion.

Assignment

1. Consider the major differences between university life for a student at the beginning of the twentieth and twenty-first centuries. What has changed, and what has remained the same?
2. What does 'academic freedom' mean to you? How important is it, and how much freedom do you have to pursue your interests?
3. What are your motivations for pursuing an academic education? Which values do you consider most important?

Suggested Reading

The Cambridge History of Sciences, vol. 7 (2003, edited by Roy Porter) offers an excellent collection of short but insightful essays on the history of the various disciplines in the social sciences. For an exploration of the philosophy of the social sciences, we recommend the accessible *Contemporary Philosophy of Social Science* by Brian Fay (Fay 2005) and for a more critical perspective, *Philosophy of the Social Sciences* by Alexander Rosenberg (Rosenberg 2008). For an introduction into theories of knowledge, we recommend Duncan Pritchard's *What Is This Thing Called Knowledge?* (Pritchard 2006).

References

Anderson, L. W., Krathwohl, D. R., Airasian, P. W., Cruikshank, K. A., Mayer, R. F., Pintrich, P. R., Raths, J., & Wittrock, M. C. (2001). *A taxonomy for learning, teaching, and assessing: A revision of Bloom's taxonomy of educational objectives*. New York: Pearson, Allyn, & Bacon.

Eraut, M. (1994). *Developing professional knowledge and competence*. London: Routledge.

Fay, B. (2005). *Contemporary philosophy of social science: A multicultural approach*. Boulder: Taylor & Francis.

Gabelnick, F. G. (1990). *Learning communities: Creating connections among students, faculty, and disciplines*. San Francisco: Jossey-Bass.

Giddens, A. (1991). *Modernity and self-identity*. London: Polity Press.

Merton, R. K. (1942). A note on science and democracy. *Journal of Legal and Political Sociology, 1*, 115–126.

Nussbaum, M. (2010). *Not for profit: Why democracy needs the humanities*. Princeton: Princeton University Press.

Porter, R. (Ed.). (2003). *The Cambridge history of science* (Vol. 4). Cambridge: Cambridge University Press.

Pritchard, D. (2006). *What is this thing called knowledge?* London: Routledge.

Repko, A. F., Szostak, R., & Buchberger, M. P. (2014). *Introduction into interdisciplinary studies*. London: Sage.

Rosenberg, A. (2008). *Philosophy of the social sciences*. Boulder: Taylor & Francis.

Wenger, E., McDermott, R., & Snyder, W. M. (2002). *Cultivating communities of practice*. Boston: Harvard Business School Press.

Wootton, D. (2015). *The invention of science*. London: Penguin.

Ziman, J. (2000). *Real science: What it is, and what it means*. Cambridge: Cambridge University Press.

References for Case Study: Max Weber and the Calling of Science

Hacket, E. J. (1990). Science as a vocation in the 1990s: The changing organizational culture of academic culture. *The Journal of Higher Education, 61*(3), 241–279. https://doi.org/10.1080/00221546.1990.11780710.

Weber, M. (1919/1991). Science as a vocation. In H. H. Gert & C. W. Mills (Eds.), *From max Weber: Essays in sociology* (pp. 129–156). London: Routledge.

Chapter 3
Perspectives

Contents

This chapter has been co-authored by Naomi van Steenbergen.

Electronic Supplementary Material: The online version of this chapter (https://doi.org/10.1007/978-3-030-48415-6_3) contains supplementary material, which is available to authorized users.

© The Author(s) 2020
J. Bos, *Research Ethics for Students in the Social Sciences*,
https://doi.org/10.1007/978-3-030-48415-6_3

After Reading This Chapter, You Will:

- Better comprehend what research ethics is about
- Understand why research ethics is an integral part of doing research in the social sciences
- Be able to distinguish between research ethics and professional ethics
- Develop a general knowledge of the three most important theories of ethics

Keywords Board of complaint · Code of conduct · Consequentialism · Deontology · Deplorable practices · Ideal practices · IRB · Integrity · Ombudsperson · Principled sensitivity · Professional ethics · Questionable research practices · Research ethics · Shared values · Ideal research behavior · Deplorable research behavior · Questionable research practices

3.1   Introduction

3.1.1   Worst Case Scenario

In August 1971, Stanford University psychology professor Philip Zimbardo was running just the first week of his grand 'Prison Experiment', and participants were already behaving according to the roles assigned to them. The twelve 'prisoners' were becoming passive, subordinate, almost inert, while the twelve 'guards' were behaving more and more like bullies.

The prisoners and guards, who were all male students at Stanford University and voluntary participants in the experiment, were assigned their respective roles randomly at the offset. Zimbardo's goal was to research the psychological effects of perceived power. In particular, he was interested in 'deindividualization' and 'dehumanization' (loss of personhood).

There was a major problem developing: the experiment was quickly getting out of hand. The 'guards' began to behave callously, and the 'prisoners' were beginning to actually suffer. After some deliberation, Zimbardo decided to discontinue the experiment.

The Stanford Prison Experiment (SPE) has been the subject of heated debate to this day (see Haslam and Reicher 2012; Haslam et al. 2019). In a reanalysis of the original data, Le Textier (2018) found that Zimbardo's narrative of the experiment was flawed in a number of respects. Le Texier found that the data presented by Zimbardo was incomplete and biased towards dramatization and did not disclose that the guards acted on precise instructions from Zimbardo, whose 'experiment' seemed to be designed more as a *demonstration* than a scientific study. Bartels and Griggs (2019) concluded from these criticisms that textbooks should revise and repurpose the coverage of the Stanford Prison Experiment. Textbook authors should use the SPE as a case to teach students the importance of critical thinking, and the value of self-correction in science.

Indeed, the educational value of historical cases such as these lies less in their isolated ethical and methodological shortcomings, of which Zimbardo seemed acutely aware. Rather, their value resides in understanding how our moral beliefs have changed over time. What are our ethical presumptions, and by which norms do we live?

Getting an answer to these questions is the purpose of this chapter. In order to answer them, we need to begin at what ethics are and why they are important. To do this, we will explore the various approaches to ethics.

3.1.2 What Are Ethics?

Peter Singer (2001), professor of bio-ethics at Princeton University, notes that people often like to believe that ethics are just 'an annoying list of things you are not allowed to do, so you can't have fun.' Rather, he says, that's not the case. Ethics are an inquiry into what is right and wrong, and what is valuable and important. It attempts to answer the question of what you *ought* to do.

When performing research, you are inevitably going to make decisions that will affect others, and you need to know which of the available options is the best course of action. Which one to choose is not always immediately clear. Here are a few examples:

- Suppose you want to know a respondent's view on a particular subject, but directly asking about it would likely influence the respondent. Is it therefore acceptable to mislead the respondents so the information you receive is more valid? Or should you be honest and open about your intentions, which would require fully informing the participant and possibly affecting your data?
- Suppose you want to investigate certain behaviors, and to do so, your respondents need to perform a task that carries a small risk about which the respondent is fully informed. Do you still have a responsibility for the safety and well-being of the respondents even if they are fully informed about the risks? If so, how far does that responsibility reach? Does it end with the experiment or should you provide care afterwards?
- Suppose you have made a discovery that could benefit some, but harm others. Should you publish the results or not? How should you reach your decision and on what criteria?

The answers to many of these questions have already been formulated, either in the form of general principles to be followed ('codes of conduct'), or in the form of very specific rules that apply to certain situations or conditions ('do's and don'ts'), to which we turn later in this chapter.

However, the strictest of rules leave room for interpretation, and even in the most clear-cut cases, there may be more than one solution available.

Thus, knowing what to do requires a degree of *principled sensitivity*, meaning researchers should be sensitive to the rights of others and their well-being. Many argue, furthermore, that this sensitivity does not stop at individuals; it stretches into communities, animals, and even the environment at large.

3.1.3 Three Cases

Before we go into detail on the ethical approaches in research, let us first examine three concrete examples that will allow us to appreciate the various dimensions of ethics in a broader sense.

First, consider the horrendous hypothermia experiments carried out by Nazi doctors on prisoners during the Second World War. Prisoners were strapped naked to a stretcher in the Polish winter or immersed in ice-cold water while data on their bodily response was meticulously collected. The Nazis used this data to determine how much cold a human body could endure, arguing that it would come in handy on the Eastern Front (Berger 1992).

About a third of the prisoners did not survive the experiments. Post-war abhorrence for these types of experiments led to the 'Nuremberg Code of Ethics', on which our present-day ethical codes are based.

The question here is obviously not whether there is an ethical dilemma about the experiment's procedures. The question is: could the data still be used? This question has been subject of an ongoing post-war debate (Schafer 1998). In a discussion on this issue, David Bogod (2004, p. 1156) contends that unethically acquired data should never be used. That being said, he does acknowledge that others would argue, 'if some general good can come of the most evil acts, then those who suffered and died might not have done so entirely in vain.' However, this quickly leads to a follow-up question: when you honor that argument, aren't you assuming that the subjects are providing a posthumous consent? The very first principle of the Nuremberg Code outlines that 'voluntary consent of the human subject is absolutely essential.'

Second, consider the well-known 1961 obedience studies conducted by Yale psychologist Stanley Milgram. Milgram (1963) led his subjects to believe they were taking part in an experiment on learning. In their role as a 'teacher', the participants were to administer electric shocks to a fellow participant, the 'learner.' The shocks supposedly increased in severity over the course of the experiment, and the response from the learner became ever more volatile. What the subject didn't know was that in reality, the learner was a stooge, an actor hired by Milgram to feign pain, never actually receiving the electric shocks.

Milgram rationalized 'obedience' as the willingness to 'carry out another person's wishes.' He wanted to know whether his research participants would continue participating in his study, and at what point would they decide that their collaboration was no longer justifiable. However, when his subjects voiced doubts and proposed to stop, the researcher answered with a line from a script: 'The experiment requires that you continue…' (see Perry 2013) (Figs. 3.1 and 3.2).

Milgram's work has been widely criticized, especially on ethical grounds. The controversy even affected his professional career (he was denied tenure at Harvard, see Miller 1986). Did his subjects know what they were in for? Had they been

Fig. 3.1 The Milgram
Experiments. Subject in
the study (c) Yale
University Manuscripts
and Archives

subjected to a tolerable level of pressure? Were they properly debriefed afterwards
and was there any form of care offered if needed? The answer to most of these questions is 'no.' For that reason, the Milgram studies, even more so than the Stanford
Prison Experiment, stand out as a landmark of unethical research.

Our third example questions research on the origin of sexual orientation, specifically homosexuality or 'same-sex sexual orientation.' Homosexuality has long
been, and in certain societies still is, illegal. This is justified because it was (or still
is) considered an aberration of the norm (Greenberg 1988).

Research into the 'origin' or 'cause' of homosexuality has by and large adopted
this perspective of 'abnormality.' Thus, psychoanalytic theorists proposed that
homosexuality may be caused by 'arrested psychosexual development,' often in the
context of a dysfunctional family constellation. Forms of psychotherapy, it was reasoned, ought to be based on the idea that it could and should be possible to shift the
sexual orientation back to 'normal' (Halderman 1994).

With the gay rights movement of the 1960s and 70s, the official view changed.
Homosexuality would no longer be considered an abnormality. Interestingly,
researchers continued to search for the cause of homosexuality, though now in
(socio)biological terms (notably genetic, hormonal, and environmental influences).
This suggested that at its base, homosexuality was still something 'unnatural' and in
need of an explanation. The underlying ethical consideration here is voiced by
Schüklenk et al. (1997, p. 10): 'Why is there a dispute as to whether homosexuality
is natural or normal? We suggest it is because many people seem to think that nature
has a prescriptive normative force such that what is deemed natural is necessarily
good and therefore *ought* to be.'

From these brief cases we extract two provisional observations:

First, *ethics in science is a complex and multifaceted issue*. It is not just about
setting up proper research protocols or treating subjects respectfully (though that is
certainly important). It is also about the types of questions asked, the (implicit)
presuppositions made, and the ways data is analyzed and communicated. This
includes the impact scientific research may have on society and the responsibilities
researchers have towards individuals and communities.

Second, *norms and values are not fixed objects*. Although there is universal consensus on certain basic values ('do not harm', for example), sensitivity to other

Public Announcement

WE WILL PAY YOU $4.00 FOR ONE HOUR OF YOUR TIME

Persons Needed for a Study of Memory

*We will pay five hundred New Haven men to help us complete a scientific study of memory and learning. The study is being done at Yale University.

*Each person who participates will be paid $4.00 (plus 50c carfare) for approximately 1 hour's time. We need you for only one hour: there are no further obligations. You may choose the time you would like to come (evenings, weekdays, or weekends).

*No special training, education, or experience is needed. We want:

Factory workers	Businessmen	Construction workers
City employees	Clerks	Salespeople
Laborers	Professional people	White-collar workers
Barbers	Telephone workers	Others

All persons must be between the ages of 20 and 50. High school and college students cannot be used.

*If you meet these qualifications, fill out the coupon below and mail it now to Professor Stanley Milgram, Department of Psychology, Yale University, New Haven. You will be notified later of the specific time and place of the study. We reserve the right to decline any application.

*You will be paid $4.00 (plus 50c carfare) as soon as you arrive at the laboratory.

TO:
PROF. STANLEY MILGRAM, DEPARTMENT OF PSYCHOLOGY, YALE UNIVERSITY, NEW HAVEN, CONN. I want to take part in this study of memory and learning. I am between the ages of 20 and 50. I will be paid $4.00 (plus 50c carfare) if I participate.

NAME (Please Print) ....

ADDRESS ....

TELEPHONE NO. ....Best time to call you ....

AGE .... OCCUPATION .... SEX ....

CAN YOU COME:

WEEKDAYS .... EVENINGS .... WEEKENDS ....

Fig. 3.2 The Milgram Experiment. Left: Advertisement in the *New Haven Register* June 18, 1961. Right: Subject in the study © Yale University Manuscripts and Archives values may change over time and can differ from society to society. For example, we have witnessed over the past several decades an increased concern with data manipulation and data storage, while the emergence of the internet has given rise to new questions regarding confidentiality. These concerns have led to stricter regulations in many countries. However, value differences mean the policies deployed to handle these considerations differ between China, the US, and many European countries, and this may complicate data sharing in the future (Box 3.1).

Box 3.1: The Ethics of Eating Disorder Research
The case detailed below, borrowed from Wassenaar and Mamotte (2012), questions the ethics of an eating disorder study conducted by a researcher from South Africa. The South African researcher in question was interested in establishing the cross-cultural validity of the Eating Disorder Inventory (EDI), developed and standardized in the United States.

The EDI consists of 64 questions, clustered into eight subscales, that all relate to the psychological conditions of anorexia nervosa and bulimia, seeking to determine the degree of their disorder. They sought to answer: does the EDI work in 'developing countries' as well?

To probe this question, the researcher aimed to have some 500 female university students fill out the EDI. They then sampled high and low scoring participants who would be subsequently interviewed by skilled clinicians on a blind basis (neither the participant nor the interviewer would know the participants' EDI score).

The aim of the study was to determine whether high and low EDI scores correlated with the clinician's estimate of the severity of the participant's eating disorder. Participants would be compensated with 'study credits' (points awarded to students for participating in research).

Though the wish to establish the cross-cultural validity of the EDI is legitimate, Wassenaar and Mamotte voiced several ethical concerns with this research project, three of which are highlighted below:

- First, the community in which the research was conducted (university campus) had not been informed of nor approved of the study. Research must respect community cultures and values, and therefore the researcher should have formed collaborative partnerships with women's health groups on campus or with class representatives before undertaking the study.
- Second, the participants had not been fully informed about the risks involved. Indeed, research on emotionally and socially sensitive topics like eating disorders can potentially induce harm, such as anxiety, painful self-discoveries, stress, indignation, and secondary traumatization. The ethical review board, whose task it is to oversee these potential dangers and who had approved of the research, underestimated the potential for emotional distress.
- Third, there is the issue of dependency. Students are in an unequal power relationship with teachers and/or researchers. They may feel that a failure to

(continued)

Box 3.1 (continued)

participate can lead to disapproval, or that they are not free to withdraw their collaboration. Wassenaar and Mamotte suggest that a structured assessment instrument be used to evaluate the participant's ability to consent to research and understand the voluntary nature of their decision to participate.

In conclusion, Wassenaar and Mamotte determined that there is a 'need for special ethics scrutiny of mental health related research proposals involving students as research participants.'

Do you agree with this conclusion? Or do you believe that students do not differ from any other research population, and thus need no special 'ethics scrutiny' when they are involved in health-related research? Why or why not?

3.2 Conceptualizing Research Ethics

3.2.1 Responsible Research Conduct

From our previous discussions, we have learned that some researchers 'play by the book'; they follow procedures and aim to make the right decisions based upon conventional wisdom. Others, however, break from these conventions and make the wrong decisions; and they risk being accused of fraud. Following Steneck (2006), we will call the research practices of the former 'ideal,' and those of the latter 'deplorable.'

Ideal research behavior takes into account existing norms, institutional standards, and international legislation. Deplorable research behavior violates these norms deliberately. These practices come in the form of Plagiarism, Falsification, and Fabrication, or PFF (see Chaps. 4, 5 and 6 in this book).

It is assumed that deplorable research behavior is rarer than ideal behaviors, but there are more forms of research behavior than just ideal and deplorable. In between these extremes there exists a rather large 'grey area,' or behaviors that are neither ideal nor deplorable, but rather 'questionable.' Questionable Research Practices (QRPs) violate the established norms, but not enough to qualify as 'fraud.'

In later chapters we explore questionable practices in greater detail. For now, we observe that these practices confront us with a challenge: we need to answer the question where to draw the line. What do we find acceptable, what do we reject, and why?

3.2.2 Research Ethics and Integrity

How do we distinguish between ideal and deplorable research practices? To answer this, we need to acknowledge another distinction, namely between *research ethics* and *integrity.*

First consider *integrity*, which we understand to be the quality of having strong moral principles, like honesty or compassion. Macrina (2005, p. 1) notes how this term raises an 'image of wholeness and soundness, even perfection.' Integrity is often used as an adjective, mainly to describe one's behavior as *being integrous*. Accordingly, research integrity can be defined as 'the quality of possessing and steadfastly adhering to high moral principles and professional standards, as outlined by professional organizations, research institutions and [...] the government and public' (quoted in Steneck 2006, p. 55). In research, questions of integrity often relate to methodological and procedural issues.

Next consider *ethics*. Compared to integrity, ethics has to do with moral principles and questions of fairness and even justice. In *research ethics*, we are concerned with moral problems related to the practice of research involving living participants (animals as well as humans, individuals as well as groups, and even entire societies). The focus here is more on protecting participants, ensuring their interests and rights, and on assessing risks and protecting confidentiality, among other issues.

Though both concepts emphasize different aspects of normative behaviors, what they have in common is:

- some concept of (contested) normative *rules;*
- some notion of *communality;*
- some sense of (individual and collective) moral *responsibility;*
- and some connection to *behavior.*

In this book, we investigate the dimensions of responsible research conduct; both the procedures and the principles, the abstract norms, and the concrete behaviors (Fig. 3.3).

Fig. 3.3 Some idea of moral responsibility

3.2.3 Research Ethics and Professional Ethics

A final distinction that needs a brief exposition is that between *research ethics* and *professional ethics*.

Research ethics has to do with norms, values, and practices concerning the collection, analysis, and dissemination of scientific findings *about* the world. *Professional ethics* has to do with norms, values, and behaviors concerning the work of a practitioner (i.e. a therapist, counselor, educator, policy maker, etc.) who intervenes *in* the world.

As a researcher, you are concerned with asking relevant questions, using validated methods, obtaining reliable data, and drawing logical conclusions. As a practitioner, you are concerned with making correct diagnoses, finding effective treatments, and measuring the effectiveness of intervention (among others).

Both roles involve different sets of (normative, or standard) rules and principles that outline desirable and undesirable behaviors. But this sharp division of labor does not always represent reality. Here are two examples where a strict interpretation of a researcher or practitioner's responsibility becomes problematic.

First, consider that researchers should be, by default, committed to the principle of anonymity, which means that data cannot be traced back to any individual. Then what should be done with unexpected findings that could be of great importance to the participant? Say, for example, that a researcher uses fMRI scans to investigate certain brain activities and by accident finds that one of the participants may have developed a tumor. Should the researcher break from the normative rules of research ethics like anonymity and assume some form of professional ethics in order to refer the participant to a specialist? Doing so would lead them to intervene *in* the world, as if they were a practitioner. Is this acceptable? Similar questions can be raised when there is suspicion of child abuse or marital violence (these questions on confidentiality are addressed in Chap. 7).

Second, consider the responsibilities of a researcher who collaborates with a third party (for example a governmental body, a professional organization, or a special interest group). While on one hand the researcher needs to maintain scientific objectivity, they may also need to take into account some of the concerns specific to that field or those of a specific organization, which can lead to conflicts of interest (this will be addressed in Chap. 8).

While this book will focus on research ethics time and again, we will find that cases overlap with aspects of professional ethics, and that in our deliberations, we cannot rely solely on research procedures to make the right call.

3.3   Codes of Conduct

3.3.1   Guiding Principles

Ethics (in a prescriptive sense) is reflection on what actions or behavior might be justified. This reflection can and often does result in normative rules or principles. Nevertheless, there is not one specific set of well-defined rules that specifies exactly which behavior qualifies as ethical.

First, a list of rules covering all possible ethical decisions across every possible situation would be endless.

Second, even if we had such a list, real life situations are complex and ambiguous, and can rarely be governed under just one principle or one rule.

Fortunately, there are *guiding* principles that can help us navigate our way through normative issues. These guiding principles are called 'codes of conduct.'

Today, all universities require that its members (staff as well as students) adhere to such a code of conduct, often modelled after similar codes first introduced in the medical professions.

Codes of conduct can differ from discipline to discipline and even from culture to culture, though all share a number of notably important principles (see Box 3.2 for a list of shared values often found in academic codes of conduct).

Box 3.2: Shared Values in Scientific Research
Accountability: Be reliable and responsible with your research, from idea to publication.
Animal Care: Show proper respect and care for animals when using them in research. Do not conduct unnecessary or poorly designed animal experiments.
Carefulness: Try to avoid careless errors and negligence. Keep good records of your research activities, research design, and correspondence with agencies or journals.
Competence: Maintain and improve your own professional competence and expertise through lifelong education and learning. Take steps to promote competence in science as a whole.
Confidentiality: Do not disclose the personal information of research subjects, nor their identities. Protect sensitive information.
Honesty: Convey information truthfully. Honor commitments. Do not fabricate, falsify, or misrepresent data. Do not deceive colleagues, research sponsors, or the public.
Human Subjects Protection: When conducting research on human subjects, minimize harm and risks, and maximize benefits. Respect human dignity,

(continued)

Box 3.2 (continued)

privacy, and autonomy. Take special precautions with vulnerable populations. Strive to distribute the benefits and burdens of research fairly.

Legality: Know and obey relevant laws, institutional codes of conduct, and governmental policies.

Non-Discrimination: Avoid discrimination of anyone on the basis of sex, race, ethnicity, or other factors not related to scientific competence and integrity.

Objectivity: Strive to be impartial and avoid bias and self-deception. Disclose personal or financial interests that may affect your research practice.

Openness: Share your data, results, ideas, tools, and resources. Be open to criticism and new ideas.

Respect for Intellectual Property: Honor patents, copyrights, and other forms of intellectual property. Do not use unpublished data, methods, or results without permission. Give proper acknowledgement when credit is due.

Responsible Publication: Publish in order to advance research and scholarship, not only to advance your own career. Avoid wasteful and duplicative publication.

Social Responsibility: Strive to promote social good and prevent or mitigate social harms through research, public education, and advocacy.

[Adapted from Shamoo A. and Resnik, D. (2015). *Responsible Conduct of Research,* 3rd ed. New York: Oxford University Press.]

In Europe, universities have largely committed to the *European Code of Conduct for Research Integrity* (3rd revised edition, published in 2017). In just a few pages, the European Code of Conduct outlines the four most basic principles researchers should adhere to: *reliability, honesty, respect,* and *accountability*. It also outlines a general sense of 'good research practices' regarding (among other things) training, supervision, and mentoring of researchers, as well as how to establish sound research procedures.

Many European countries have created their own codes of conduct on top of the this code, defining in somewhat greater detail what is required of their researchers. For example, every university in the Netherlands has accepted the *Netherlands Code of Conduct for Scientific Integrity* (last updated in 2018), a 30 page document that details its founding principles, specifies the norms of 'good scientific research practices,' and stipulates the universities' obligations with respect to issues such as training, supervision, data management, and procedures regarding scientific misconduct. Most universities, though not all, have created a special student version of their code of conduct designed to outline proper behavior in class and on campus.

Promoting ethical policies and codes of conduct has become a major task of specialized bodies within institutes (see Iverson et al. 2003). In compliance with international regulations, most universities have established special Institutional Review Boards (IRBs) to safeguard 'research ethics,' about which we write in greater detail in Chap. 10. Furthermore, they have assigned ombudspersons (inter-mediaries between an institution and the government) and boards of complaints designed to safeguard 'research integrity.'

Often these specialized bodies have statutory and disciplinary powers, meaning they can and will take action in case ethical integrity has been violated (rights of participants, or research subjects). Failing to comply with a code of conduct may indeed carry disciplinary action against the offender, differing from an official slap on the wrist or being put on probation, to even discharge from one's position (staff member) or removal from the institute (student). Note how ethical consideration thus carry legal consequences. Although this aspect of research ethics is not elabo-rated upon further in this book, it is an important reality to carry with you. In the last chapter of this book though, we will return to the task of IRBs.

3.3.2 Key Imperatives

Codes of conduct generally outline the subjects or issues that we should pay attention to. They do not give us precise guidelines, as the task of ethics is not to specify exactly which behaviors are desirable and which are not. However, there are a few exceptions – certain behaviors that the scientific community agrees are desir-able (and where the opposite behavior is undesirable). These are called the *impera-tives of science*. An imperative is a rule or principle considered to be crucial or decisive. It tells you where to draw the line. In the social sciences (and indeed in science more generally), the following imperatives have been universally accepted as fundamental to the practice of research and can be found referenced in any text-book on ethics:

- *Avoid harm and do good.* Researchers have an obligation to improve, promote, and protect the health of people and their communities. They must furthermore seek to avoid any harm done to human participants, or to animals, and must seek to minimize the risk thereof.
- *Respect for persons.* Researchers must protect the autonomy of research partici-pants. This imperative implies recognition of persons as autonomous, unique, and free subjects. It also means that researchers acknowledge that each person has the right and capacity to make their own decisions, including the right of non-participation.
- *Protect confidentiality.* Participants must be sure that their data is processed anonymously (unless there is a reason not to, and the participant is notified thereof). No participant should suffer consequences from having participated in any research because certain personal information is made public.

- *Avoid deception*. Participants may not be deceived, misinformed, or misled by researchers (unless there is good reason to, and the deception is debriefed afterwards). In line with the imperative of autonomy, human research participants should be considered capable of deciding whether they consent and to what it is they are consenting (Box 3.3).

Box 3.3: Bothersome Research: A Dilemma

You are conducting clinical research for which you need a lot of patients. Some of the patients are very ill and it becomes clear they would prefer to not participate in the research at all. You respect this and conduct the research with healthier participants. After all, you deduce, there is a certain amount of stress involved without evidence of benefits. A couple of days later, you receive an email from your professor in which he makes it clear that you are behind schedule and should collect the data of at least ten new patients before the end of the week. This would mean that you must include the very ill patients, despite their wish to not be included. Things are not going well with your professor because you failed to come up with any significant results in your last research project. What do you do?

(a) Go back and thoroughly explain the importance of the research to the patients, asking again if they would participate.
(b) Explain the patients' situation to the professor and emphasize that they do have the right to refuse to participate.
(c) Ask the professor to extend the period of data collection so you have time to search for other patients. You know he will not be pleased with the request.
(d) Discuss the issue with the medical personnel, and request that they ask the patients again on your behalf.

[Case adapted with permission from *Dilemma Game: Professionalism and Integrity*, Erasmus University Rotterdam].

3.4 Fundamental Dilemmas and Ethical Theories

3.4.1 The Need for Ethical Reflection

Thus far, we have discussed normative rules and moral principles, some shared values and ethical norms, and various imperatives that are intended to guide scientific research. But how should we understand the role of these norms, rules, and principles? Are they set in stone? Where do they derive from? What if they clash or fail to give adequate guidance?

For a start, it is important to recognize that even in a discipline that is governed by widely accepted norms and principles regarding research, researchers may find themselves confronted with ethical dilemmas. What, for instance, about the imperative to avoid harm and do good? These are in fact two imperatives that in certain cases may prove mutually exclusive. Sometimes, it seems, some (potential) harm must be done to a research subject in order to do good overall.

For example, an intensive care unit at an academic hospital may drastically improve the quality of its care and the survival rate of its patients by gathering extensive data from critically ill patients. It is not, however, able to ask permission from these patients and therefore gathering and using the data violates the usual norms and guidelines regulating consent. Or what about a government-run statistical service, that may, by gathering traffic data, be able to help improve the efficiency and safety of the transport infrastructure. Here, too, the people whose data is gathered could not possibly be asked for consent, which the statistical service's code of conduct requires. Does the (potential) good achieved in these cases outweigh the harm done by gathering and using people's data without consent?

Even more complex questions may arise. What, for instance, should we consider harm in the first place? Being manipulated to make a different choice than one would have otherwise made goes against autonomy, which many consider an important value. But what if this alternative option is objectively better for the research subject concerned, for instance because it is healthier or more cost-effective? Does the manipulation still count as harm? Or is it actually an example of doing good?

As these brief examples show, even when clear norms and principles are available for a research field, ethical questions still arise. Moreover, beyond ethical dilemmas stemming from incompatible or insufficiently clear guidelines, ethical reflection on norms and principles is important because we cannot blindly assume that all existing guidelines will always remain, or indeed are currently, ethically justified. Standards from the past have been revised in the light of new insights or developments and there is no reason to think that our current standards will suffice for the decades to come.

For all of these reasons, it will be important for any researcher to be able to connect critically to the norms and principles of their own discipline. But where can someone looking for ethical guidance beyond established norms turn? This question pushes us deeper into the domain of ethics in the sense of reflection on what actions might be justified. Obviously, the scope of this chapter does not allow for an in-depth discussion on ethical theories. However, a first sense of some of the approaches that inform research codes of conduct, norms, and principles might be useful.

3.4.2 Deontology Versus Consequentialism

Many principles in research codes of conduct follow (in terms of ethical theory) a *deontological* approach. In order to understand what this approach entails, it is helpful to contrast it with its main rival: *consequentialism*. According to

consequentialists, we must judge rules or particular actions by the specific consequences of these rules or actions. Many consequentialists in addition hold that the
way in which we ought to judge consequences is by the question of how much good
(or well-being) the action or rule would produce for all involved parties combined.
In other words: the best (or even the only justifiable) action or rule would be the one
that results in the most good overall. A consequentialist might therefore, to give an
extreme example, judge that the right thing to do is to sacrifice one person in order
to save or help many (assuming that this sacrifice would indeed result in the most
good overall).

In contrast, one of the core convictions of deontologists is that there are certain
things we must always or may never do, regardless of whether deviating from these
norms might have the best result in a particular situation. This conviction is something that is quite easily recognizable in many of the norms, rules, and imperatives
discussed above. Informed consent must *always* be obtained, regardless of whether
your results might be better if you didn't. Research subjects must *always* be
debriefed, even if this makes it harder to do a second run of the same experiment.
You may *never* break confidentiality, even if disclosing the personal data of your
research participants to a third party would enable you to generate extremely interesting results.

The use of 'always' and 'never' conveys a close alignment with the deontological approach. You might think, however, that the fact that research codes of conduct
are (superficially) deontological in character ultimately is (or should be) the result
of a consequentialist kind of reasoning: if every individual researcher were to set
their own rules based on their own judgement, this would lead to a mess, public
distrust in science, or other unfavorable results. The conviction that a discipline
ought to abide by strict norms and imperatives for its research conduct can therefore
be motivated by either deontological or consequentialist approaches.

Furthermore, you might think that strict adherence to a set of imperatives is neither helpful nor desirable. You may, for instance, wonder whether it may sometimes
be right to forgo informed consent if the expected consequences of doing so promise to be very good and the harm seems relatively small (such as in the example of
the intensive care unit above). This is the sort of reflection that can be helped by a
better grounding in ethical theory.

Both consequentialism and deontology have a wide following amongst ethicists.
Important to remember for the purposes of this chapter is that there is not only significant discussion on whether our ethical decisions should follow deontological or
consequentialist principles, but also on which actions these sets of principles would
call for. Should deceiving a research participant in all forms and circumstances be
prohibited, according to deontology? It's an open question. Should a consequentialist accept medical interventions on a test subject against their will if this is very
likely to improve the life expectancy of many others? It's debatable. These theories
are helpful instruments in considering such questions, but they do not settle the
questions definitively in any simple way.

3.4.3 Virtue Ethics

Finally, it is worth looking briefly at a third approach, known as *virtue ethics*. Virtue ethicists offer a different perspective on research ethics, for they might say that instead of relying too much on codes of conduct and lists of rules, it is important for researchers to cultivate relevant virtues such as honesty, reliability, humility, and conscientiousness. There are different reasons for this, but a central one is that, as we have seen, norms and imperatives can clash and often do not determine exactly what is required in any given situation. For this reason, virtue ethics places a strong emphasis on judgement and so-called 'practical wisdom,' which allows one to decide exactly what action the set of virtues calls for in a specific context. So, unlike a researcher merely abiding by lists of rules who is stumped whenever that list does not suggest a clear and definitive course of action, a virtuous and practically wise researcher would be able to judge what their virtues require of them in specific situations. The downside of this approach is that cultivating such virtues and the practical wisdom to apply them is not an easy thing to do – it is a long and difficult process that involves a lot of practice. Virtue ethicists correspondingly would attach great importance to education in these virtues and good examples being set by other members of the profession.

3.4.4 Ready-made Solutions?

From our discussion, it has become clear that ethical questions may arise in any research context and that codes of conduct, lists of values, or imperatives do not provide any ready-made answers. While ethical theories can help to reflect on the arguments and considerations underlying possible courses of action, they do not solve the ethical questions all by themselves. Codes, rules, and procedures provide us with indications, and in some cases strong indications of what to do. However, at the end of the day, there rests a moral responsibility on the shoulders of the researcher to justify their actions (which choices they made and on which grounds) and explain them to others (other researchers, participants, the community). This task becomes increasingly more important when society demands accountability.

3.5 Conclusions

3.5.1 Summary

In this chapter, we have familiarized ourselves with research ethics in a general sense. We have come to understand it as *the application of normative rules or principles, such that you know how to behave in a responsible way.* We distinguished between 'ideal' and 'deplorable' practices in research and differentiated between research ethics proper (which applies to norms, specifically with regard to working with living participants) and research integrity (which is defined as maintaining high moral principles and professional standards).

We have learned, furthermore, that research ethics require *sensitivity and responsiveness* on the part of the researcher, who must be wary that the ethical dimensions of their work can be *highly contested.*

We have also established that ethics implies an *obligation* on the part of the academic community to provide the guiding principles and institutional imperatives that allow research to take place at all. These guiding principles translate into specific *codes of conduct* that aim to formalize desirable behavior and prevent undesirable behavior.

Finally, we discussed the unavoidability of reflecting on ethical questions surrounding research conduct. We established that neither codes of conduct, disciplinary norms, or broadly shared imperatives, nor fundamental ethical theories provide ready-made answers. Both leave us with the responsibility to develop our own stance. In short, and returning to where we started, there is a profound message in the observation that ethics pose a significant challenge, for which we must prepare ourselves.

3.5.2   Discussion

Two issues have remained unresolved in this chapter. One is the ratio between individual responsibility and institutional responsibility. Where does your personal responsibility end, and where does that of your institution begin? The other is practical. *How* do you develop this sense of responsibility? We hope to help you answer these two questions in the remaining chapters of this book, when we discuss the most important ethical considerations of research practices step by step.

Case Study: The Ethics in Suicide Prevention Research

In 2014, the World Health Organization (WHO) called for a worldwide escalation of suicide prevention efforts, with further systematic research into the effectiveness of suicide prevention therapies deemed necessary. However, research into such therapies is often hampered by the fact that individuals perceived to be at risk of suicide are, as a rule, excluded from participation in research on the grounds that death or self-injury may occur during trials.

Perhaps the most pressing dilemma in the study of suicide prevention is how to ensure that 'good quality, ethically sound research' is fostered in a way such that 'we can better understand, appropriately respond to, and reduce the incidence of suicide' (Fischer et al. 2002, p. 9).

Fig. 3.4 Respecting others and offering help

There are a number of questions that should be taken into consideration: (a) as researchers have an obligation to 'do good and prevent harm,' they must thoroughly assess any risks for the participants involved, and therefore also consider liability and responsibility (both of the researcher and the institution); (b) furthermore, with regard to respecting participants' autonomy, great care must be taken that consent and confidentiality are guaranteed, and that a participant's beliefs (religious, philosophical, and otherwise) are respected; and (c) researchers have to carefully distinguish the role of researchers from that of care giver, and ensure that participants do not confuse the two (Fig. 3.4).

It is the task of ethics committees, institutional review board (IRBs), and other legislative bodies to examine research applications, and to determine whether or not an application meets their minimum requirements. In approving or denying such research, these bodies will rely on a series of procedures that check for a number of 'core principles' outlined elsewhere in this chapter.

These procedures provide general guidance in making assessments. In practice though, they cannot prevent that the same situation is assessed differently by different members of these bodies. There is, after all, not a single generally accepted moral truth to refer to (Hom et al. 2017).

How do different IRBs regard different risks, possible harm, and other ethical issues in these research proposals? Lakeman and FitzGerald (2009) surveyed 125 members sitting on human research ethics committees across five Western countries, and queried them about their experiences in assessing suicide prevention research protocols. From this survey we draw several 'areas of contention' which we grouped into three categories:

Vulnerability A major concern in suicide prevention research is that respondents might be susceptible to becoming *more* suicidal as a result of bringing attention to their suicidal thoughts and feelings. It may be harmful to discuss distressing material, or inadvertently confirm the insolubility of their problems. Therefore, some feel the need to protect vulnerable participants from themselves, especially because people with severe depression may be perceived as not competent enough to participate. In their distress, it is sometimes believed, they may not comprehend what is asked of them or do not know they can refuse cooperation. Both are essential conditions of research with human participants. Others, however, argue that assuming disability on these grounds is a violation of their autonomy; it means denying them the right to self-determination.

When confronted with this dilemma, several ethical committee members pointed out that ethics committees have a tendency to be too conservative in this respect. They expressed the opinion that it is unlikely that a researcher would make someone suicidal by merely allowing them to talk. One committee member stated, 'In my experience, my fellow members of the committee can be too cautious and protective of these subjects, who may often be very keen and willing to talk to a researcher.' However, another ethicist with many years of experience disagreed, stating: 'these people need treatment first and foremost, not study' (Lakeman and FitzGerald 2009, p. 15).

Responsibility Researchers who collect data on a vulnerable population have to make an assessment whether the potential research benefits (furthering knowledge, possible self-knowledge of the participant) outweigh the cost of intrusion on the participant. Should the answer be 'yes,' then a second concern should be raised. If during the course of research, problems are uncovered that cannot adequately be dealt with by mainstream services, then the researcher may develop a 'duty of care' to the participant, to provide or facilitate access to help' (p.16). Does the researcher have a dual role here? Do they have to provide care if necessary, or at least be adequately trained to work with clinically depressed people?

How do ethics committee members see this question? One member observed that conducting research 'increases the onus on the researcher to be able to source/provide an appropriate suicide risk assessment, safety plan, and treatment as necessary' (p. 15). Others suggested that 'not having the resources to provide help to people is a serious ethical problem particularly in terms of raising people's expectations that help will be provided' (p. 16). This calls forth yet another question. Are there any hazards involved for the researcher themselves? If the researcher is not adequately trained or does not receive adequate support, is there a risk of harm in the form of distress, guilt, or even liability?

Confidentiality Research with human participants takes place on the condition of anonymity, which means that the privacy of the respondent must be respected and that their identity remains undisclosed. This can pose a problem when a participant expresses suicidal thoughts, but the researcher has assured full confidentiality. Should the researcher notify a professional regardless? Should family members be alerted? How do ethics committee members look at these questions?

Respecting privacy on the one hand, while also offering help or reporting suicidal ideation poses particular dilemmas, according to Lakeman and Fitzgerald (2009, p. 16). Should family members be alerted in case of impending suicidal behaviors, or should the principle of privacy prevail? One of the respondents stated that family members are not informed by health care professionals. Exemplifying the varying degrees of this debate, another member stated that 'Whatever moral beliefs the researcher has about suicide should be suspended and it must be respected that the suicidal person has the right to take their own life if they so wish' (pp.17–18).

Assignment

1. Where do you stand on the issues of vulnerability, responsibility, and confidentiality? What are your ethical considerations?
2. Among the recommendations given for ethical research with people who are suicidal, Lakeman and FitzGerald (2009, p. 18) propose that researchers 'provide full information to participants about the consequences of their participation and the boundaries of confidentiality.' As a researcher, what would you consider an acceptable 'boundary of confidentiality'?

Suggested Reading

Thorough introductions into ethics are offered by Torbjörn Tännsjö, *Understandig Ethics* (2004) and Russ Shafer-Landau, *The Fundamentals of Ethics* (2015). David Resnik's *The Ethics of Science: An Introduction* (2005) stands out as a classic introduction into research ethics. For reference works, we recommend the *Handbook of Research Ethics and Scientific Integrity* (2019) edited by Ron Iphofen, and the *Handbook of Ethical Theory* (Copp 2007). Furthermore, the 2 volume *APA Handbook of Ethics in Psychology* (Knapp et al. 2012) and the *Handbook of Social Research Ethics* (Mertens and Ginsberg 2009), *Research Ethics and Integrity for Social Scientists* (Israel 2014), and the *European Textbook on Ethics* (2010), which deals with practical issues such as vulnerability, privacy, and justice. Max Weber's 1917 address on 'Science as a Vocation' remains very readable and offers valuable insights into what it means to become a scientist.

References

Bartels, J. M., & Griggs, R. A. (2019). Using new revelations about the Stanford prison experiment to address APA undergraduate psychology major learning outcomes. *Scholarship of Teaching and Learning in Psychology, 5*(4), 298–304. https://doi.org/10.1037/stl0000163.

Berger, R. L. (1992). Nazi science. In A. L. Caplan (Ed.), *When medicine went mad: Bioethics and the holocaust* (pp. 109–133). New York: Springer.

Bogod, D. (2004). The Nazi hypothermia experiments: Forbidden data? *Anaesthesia, 59*(12), 1155–1156. https://doi.org/10.1111/j.1365-2044.2004.04034.x.

Copp, D. (2007). *Handbook of ethical theory*. Oxford: Oxford University Press.

European Code of Conduct for Research Integrity. Revised edition (2017). Berlin: ALLEA – All European Academies.

Greenberg, D. F. (1988). *The construction of homosexuality*. Chicago: The University of Chicago Press.

Halderman, D. C. (1994). The practice and ethics of sexual orientation conversion therapy. *Journal of Consulting and Clinical Psychology, 62*(2), 221–227.

Haslam, S. A., & Reicher, S. (2012). Tyranny: Revisiting Zimbardo's Stanford prison experiment. In J. R. Smith & S. A. Haslam (Eds.), *Social psychology: Revisiting the classic studies* (pp. 126–141). Thousand Oaks: Sage.

Haslam, S. A., Reicher, S. D., & Van Bavel, J. J. (2019). Rethinking the nature of cruelty: The role of identity leadership in the Stanford prison experiment. *American Psychologist.* https://doi.org/10.1037/amp0000443.

Iphofen, R. (Ed.). (2019). *Handbook of research ethics and scientific integrity*. Cham: Springer. https://doi.org/10.1007/978-3-319-76040-7.

Iverson, M., Frankel, M. S., & Siang, S. (2003). Scientific societies and research integrity: What are they doing and how well are they doing it? *Science and Engineering Ethics, 9*, 141–158. https://doi.org/10.1007/s11948-003-0002-4.

Knapp, S. J., Gottlieb, M. C., Handelsman, M. M., & Vande Creek, L. D. (Eds.). (2012). *APA handbook of ethics in psychology, Vol. 1. Moral foundations and common themes.* Vol. 2. Practice, teaching, and research. American Psychological Association.

Le Textier, T. (2018). *Histoire d'un mensonge: Enquête sur l'expérience de Stanford*. Paris: Edition la Découverte.

Macrina, F. L. (2005). *Scientific integrity. Texts and cases in responsible conduct of research.* Washington, DC: ASM Press.

Mertens, D. M., & Ginsberg, P. E. (2009). *The handbook of social research ethics*. London: Sage. https://doi.org/10.4135/9781483348971.

Milgram, S. (1963). Behavioral study of obedience. *Journal of Abnormal and Social Psychology, 67*, 371–378. https://doi.org/10.1037/h0040525.

Miller, A. G. (1986). *The obedience experiments: A case study of controversy in the social sciences.* New York: Prager.

Netherlands Code of Conduct for Research Integrity. (2018). Retrieved from: https://doi.org/10.17026/dans-2cj-nvwu

Perry, G. (2013). *Behind the shock machine*. New York: The New Press.

Resnik, D. B. (2005). *The ethics of science: An introduction*. London: Routledge.

Schafer, A. (1998). On using Nazi data: The case against. *Dialogue, XXV*, 413–419. https://doi.org/10.1017/S0012217300020862.

Schüklenk, Y. U., Stein, E., Kerin, J., & Byrne, W. (1997). The ethics of genetic research on sexual orientation. *Hastings Center Report, 27*(4), 6–13. https://doi.org/10.1016/S0968-8080(98)90017-9.

Shafer-Landau, R. (2015). *The fundamentals of ethics*. New York: Oxford University Press.

Shamoo, A. & Resnik, D. (2015). *Responsible conduct of research*, 3rdrd ed. New York: Oxford University Press.

Singer, P. (2001). *Writings on an ethical life*. London: Harper Perennial.

Steneck, N. H. (2006). Fostering integrity in research: Definitions, current knowledge, and future directions. *Science and Engineering Ethics, 12,* 53–74. https://doi.org/10.1007/PL00022268.

Tännsjö, T. (2004). *Understanding ethics. An introduction to moral theory.* Edinburg: Edinburgh University Press.

Wassenaar, D.R., & Mamotte, N. (2012). The use of students as participants in a study of eating disorders in a developing country: Case study in the ethics of mental health research. The Journal of Nervous and Mental Disease, 200(3), 265–270 (2012). doi: https://doi.org/10.1097/NMD.0b013e318247d262.

Weber, M. (1919/1991). Science as a vocation. In H. H. Gert & C. W. Mills (Eds.), *From max Weber: Essays in sociology* (pp. 129–156). London: Routledge.

References for Case Study: The Ethics in Suicide Prevention Research

Fisher, C. B., Pearson, J. L., Kim, S., & Reynolds, C. F. (2002). Ethical issues in including suicidal individuals in clinical research. *IRB: Ethics & Human Research, 24*(5), 9–14. https://doi.org/10.2307/3563804.

Hom, M. A., Podlogar, M. C., Stanley, I. H., & Joiner, T. T. (2017). Ethical issues and practical challenges in suicide research. Collaboration with institutional review boards. *Crisis, 38,* 107–114. https://doi.org/10.1027/0227-5910/a000415.

Lakeman, R., & FitzGerald, M. (2009). The ethics of suicide research. The views of ethics committee members. *Crisis, 30*(1), 13–19. https://doi.org/10.1027/0227-5910.30.1.13.

Part II
Ethics and Misconduct

Chapter 4
Plagiarism

Contents

Electronic Supplementary Material: The online version of this chapter (https://doi.org/
10.1007/978-3-030-48415-6_4) contains supplementary material, which is available to authorized users.

© The Author(s) 2020
J. Bos, *Research Ethics for Students in the Social Sciences*,
https://doi.org/10.1007/978-3-030-48415-6_4

After Reading This Chapter, You Will:

- Know precisely what plagiarism is
- Understand why plagiarism is an ethical issue
- Be able to identify different forms of plagiarism
- Develop the capacity to distinguish legitimate from fraudulent references

Keywords Appropriation · Authorship · Copy-pasting · Critiquing · Inadvertent plagiarism · Paraphrasing · Patch writing · Priority disputes · Self-plagiarism · Quality control · Quotation · Summarizing

4.1   Introduction

4.1.1   The Unoriginal Sin

René Diekstra was a psychologist and tenured professor at Leiden University in the Netherlands, and a celebrated author of many popular science books. In 1996, a weekly magazine unearthed details of plagiarism in one of his books. This revelation prompted an official investigation, and as a result, he had to step down from his position (see case study for further discussion). A decade later, two German ministers (Annette Schavan and Karl-Theodor zu Guttenberg), both of whom held doctorate degree, were accused of plagiarism within a few years of one another. After an investigation into the matter, both were stripped of their academic title and had to hand in their notice (*The Guardian*, 2. 9. 2013).

High-profile cases of plagiarism with dramatic consequences such as these are by no means exceptions, nor is plagiarism a recent phenomenon. Some of the earliest reported cases of plagiarism go all the way back to the beginning of the Enlightenment. That is, back to the birth of modern science itself, and involved a few recognizable individuals, namely Newton, Leibnitz, and Erasmus (see Wootton 2015).

Though not exceptional, do these cases point to an underlying structural problem? While its prevalence is difficult to estimate, plagiarism has been found among tenured professors, but especially among students (Walker 2010), and some believe it's a rapidly growing problem. Neil Selwyn (2008, p. 468) found that nearly three in five students admit to copying a few unattributed sentences into an essay or

Fig. 4.1 Plagerius' Dilemma

assignment. Even further, one in three concede to copy-pasting a few paragraphs and just over one in ten to 'borrowing' upwards of a few pages.

Of course, there are vast differences between students, between disciplines, and between cultures (more about that later). Regardless of these differences, plagiarism in academia is an issue that cannot be chalked up to 'cultural differences' and deserves careful scrutiny. In this chapter, we explore the problem in more detail. What exactly is plagiarism, and how do we distinguish it from legitimate uses of reference literature? How does it affect our work, and which consequences does it have? Finally, we ask what factors contribute to its continued occurrence? (Fig. 4.1).

4.2 Plagiarius' Crime

4.2.1 A Working Definition

'Plagiarism' derives from the Latin noun 'plagarius,' meaning kidnapper. Plagiarism is understood as *literary theft,* namely the act of appropriating the work (or ideas) of others and passing it off as your own. As such, it stands apart from the appropriate uses of other people's work, which includes the *discussion* or *critique* of certain viewpoints, *summarizing* and *paraphrasing* of particular ideas, and the use of *quotations.* These all require the original source or author to be clearly identified. When plagiarism takes place, this is not the case (see Box 4.1).

Box 4.1: Spot the Plagiarizer!
In *The Cultural Nature of Human Development*, Barbara Rogoff (2003, p. 183) writes: 'Worldwide, child rearing is more often done by women and girls than by men and boys (Weisner 1997; Whiting and Edwards 1988).'

Here are several examples of students referencing the passage above:

Student 1: According to Rogoff (2003, p. 183), childrearing is done mostly by women and girls.

Student 2: All over the globe, childrearing is often done by women and girls and not by men and boys (Weisner 1997; Whiting and Edwards 1988).

Student 3: According to Rogoff (2003, p. 183), 'childrearing is more often done by women and girls,' but her evidence is slim.

Student 4: Many believe that childrearing is a matter for women.

Student 1 *Paraphrase*: This statement contains a clear reference to the original source (correctly identified) but it is not a direct quote, hence no quotation marks are needed.

Student 2 *Patch-writing*: The wording does not exactly follow the original, but the structure of the sentence is almost identical to it, and the references to Rogoff's sources suggest the statement is based on this literature, rather than on Rogoff. Without reference to the original source (Rogoff), this sample borders on plagiarism.

Student 3 First *quotes* then *critiques*: Quotation marks are in order here as well as a clear reference to the original source.

Student 4 Gives a *general opinion:* This opinion needs no references; it could be said by anybody.

4.2.2   Why Is Plagiarism a Problem?

Before we go into any further detail, let us first ask why it matters if you 'borrow' (appropriate) a few well-worded sentences from a source rather than crafting your own. Apart from issues of legality (making money off someone else's work), there are two moral problems attached to plagiarism: (1) Taking credit for work you have not done is deceitful; (2) science's reputation is built upon trust and accountability. Plagiarism violates the first principle and undermines the second.

There is another reason why it matters. Plagiarism may be both the smallest and the most unprofessional form of scientific misconduct, but in the eyes of the public, it is often met with more indignation than greater forms of fraud, such as data falsifying (about which we write in the next chapter). Possibly this is because it is such a noticeable form of misconduct that stands as a sharp contradiction to the high aspirations of science.

4.2.3 What Does Appropriation Entail?

By definition, whenever an author inserts a certain amount of text they didn't write themselves, and they don't adequately acknowledge the original source, that author has committed plagiarism.

This may seem clear enough, but several questions remain unanswered. When is a source acknowledged *adequately*? Is it sufficient to simply add a reference to the original text in the bibliography or is there more to it than that? Does copying just one sentence count as plagiarism? What about just *half* a sentence? In other words, is there a certain threshold after which copy-pasting counts as plagiarism? And what about translations? Suppose the original text is written in one language and you use your own translation – is that plagiarism too? And what about situations where you don't copy the original source exactly, but your work closely resembles the source in terms of structure, following the line of argumentation step by step – is that considered plagiarism as well?

In general, the answer to all these questions is: Yes – that counts as plagiarism. However, we will return to these issues in greater detail later in this chapter. For now, we suggest that you should view every text you write as a complex, layered structure, consisting of a mixture of voices: your own voice (your arguments and interpretations) and those of the people you've referenced (their arguments and interpretations). The whole point being, any reader of your text must know whom they are hearing from at all times. Every time the voice of someone else is borrowed, used, commented on, or invoked in your work, it is imperative that it be identified properly.

In any given discipline, there are specific methods for how to credit your sources. The social sciences often use the American Psychological Association (APA) citation format, whereas the humanities tend to use the Modern Language Association (MLA) format. There is no space here to discuss these methods in any detail, but we encourage you to further familiarize yourself with them (for example, see the *Publication Manual of the American Psychological Association* 2010, or the *American Sociological Association Style Guide* 2014).

4.2.4 The Question of Authorship

Failing to acknowledge referenced material is serious misconduct that can create even legal consequences when copyrights are infringed upon. This is not to imply that plagiarism can only occur with texts that are protected by a copyright, it can occur in any situation. The unacknowledged use of texts that are not under copyright, or the use of documents that have not even been published can similarly pose problems (see Saunders 2010).

Here is an example. Suppose you talk to a friend about the content of a paper you have written but not yet published, and that friend in turn uses some of your ideas in her

Box 4.2: 'Pawn Sacrifice'
Trick employed by an advanced plagiarist. A (smaller) part of a text is referenced correctly. However, a larger part of the same text is subsequently plagiarized (that is: used *without* acknowledgement). Benjamin Lahusen, in an article entitled 'Goldene Zeiten' [Golden Times] gives several examples thereof, employed in a 2005 legal textbook. The plagiarist employed the trick to mask his actions and give credibility to the rest of the text as his own (2006, p. 405).

work. Perhaps they even get their paper published before you. Would you consider that fair use? Or imagine a situation where a researcher works with a paid assistant who produces a text for them (say a short note on a particular issue). If the researcher includes these texts (in whole or in part) in their work, must they attribute each and every sentence to the assistant? Should the assistant even be recognized as a co-author of the work? What about a situation when multiple students work on a document collectively and then proceed to use portions of it in their individual papers? When they upload their work into the digital course environment, they may find that the plagiarism detection software marks their papers as being plagiarisms of one another!

Part of this problem can be resolved by referring to official (institutional) guidelines. But part of the problem surrounds group work itself, and in some of these cases plagiarism cannot be resolved simply by referring to official guidelines. Then, a resolution depends on specific arrangements being made between the parties (researcher and assistant, teacher and students, etc.) as to whom takes credit and how the work will be cited (Box 4.2).

4.2.5   When Do Intentions Come into Play?

Everybody knows it's wrong to copy-paste a certain amount of text and present it as their own. At the same time, honest mistakes happen. Tenured academics and students alike collect literature during their research, they print out articles that they believe are interesting, and they make notes on them before they start writing. Sometimes slip ups occur.

Imagine you are writing an article, the deadline is rapidly approaching, and in the final stretch you unintentionally neglect to cite a number of references. Will that be considered plagiarism? It was never your intention to plagiarize, of course!

Intentions do play a role in ethics, and they will often be factored in if plagiarism is suspected, but intentions (either good *or* bad) are difficult to prove, and good intentions do not absolve you from your duty to ascribe proper credit.

4.3   Copy-Paste Much Eh?

4.3.1   Patch-Writing

Earlier, we asked what it means to 'appropriate' a text. We found it entails the acknowledgment of the original source, and suggested that you 'identify the voice of the other.' Let us explore this a bit further. At which point do you still have to reference someone else's voice? At what point does it becomes *your* voice?

An instructive example is given by Rebekka Moore Howard in *Standing in the Shadow of Giants: Plagiarists, Authors, Collaborators* (1999, pp. 4–5). During an undergraduate course, Howard asked her students to read an excerpt from a particular source and reflect on it. In nine papers (out of a total of twenty-six) she found several sentences that had very similar wording. Here are three examples:

1. 'Specifically, "story myths" are not told for their entertainment value, rather they serve to answer questions people ask about life, about society and about the world in which they live.'
2. 'Story myths provide answers to philosophical questions about life, society and the world.'
3. 'Davidson explains that story myths provide answers to questions people ask about life, about society and about the world in which they live.'

All three of these sentences lead back to the same source, which goes as follows: 'Such "story myths" are not told for their entertainment value. They provide answers to questions people ask about life, about society and about the world in which they live' (Davidson quoted in Howard 1999, p. 5).

Howard's assignment was for her students to use *this* particular source, and these students had clearly failed to properly identify or acknowledge the original author (although some included a footnote with a reference to it). Howard's verdict was strict: all nine students received an 'F'. They were lectured on proper citation and documentation and subsequently had to revise their paper.

Commenting on this case, Howard observed how students tried to cut corners by appropriating phrases and even whole sentences from the original source. In the process, they had deleted what they considered 'irrelevant' and inserted whatever they though was appropriate. They even changed the grammar and syntax of the original sentence, 'substituting synonyms straight from *Roget's Thesaurus*' (1999, p. 6). But she also noted that, to a certain degree at least, 'patch-writing' is a matter of *style*. Even renowned scholars have been observed using 'patchwork methods,' albeit in a more sophisticated manner. In fact, the very sentence on patch-writing in this paragraph could be considered an example of patch-writing! This is all to say that the parameters distinguishing 'plagiarism' from questionable forms of 'borrowing texts' move on a sliding scale.

4.3.2   Translations as Plagiarism

Translations are not always recognized as sources of plagiarism. Consider the case below (observed by the author of this chapter). A student was caught plagiarizing when they translated a number of paragraphs originally written in English into Dutch without providing any references. Note that in the excerpt below, taken from the case, the sentence syntax and grammar are slightly altered. Furthermore, a crucial concept ('predisposition') is mistranslated (as 'factor'), and the meaning of the sentence became vaguer when it was cut into two (what does 'this' refer to in the second sentence?). In spite of all this, the original is still clearly identifiable:

Original biological predisposition is not ignored, but the focus is placed on an individual's development through interaction with other people in a certain cultural context.

Student version (Dutch in original, translated back into English) biological factors play a role, but the focus is placed on individual development. This is shaped through interaction with other people in a cultural context.
[*Biologische factoren spelen een rol, maar de focus wordt gelegd op de individuele ontwikkeling. Deze krijgt haar vorm door interactie met andere mensen binnen een culturele context.*]

As in the above case, 'translation plagiarism' appears often in a 'high-to-low form,' meaning a text originally published in a 'dominant language' (such as English) is translated in whole or in part into a 'smaller language' (say Dutch, Polish, or Italian), where it is presented as original.

Another example is found in the work of psychologist Alphons Chorus, whose *Foundations of Social Psychology* (orig. *Grondslagen der sociale psychologie*), published in 1953, contained a large number of passages lifted from a well-known introduction into social psychology by Kretch and Crutchfield, published only 5 years earlier. Though Chorus wrote in the introduction of his book that he had 'relied on Kretch and Cruchfield,' he had actually translated numerous passages word for word without providing quotation marks or references to the original. When two colleagues confronted Chorus about this, he admitted that his referencing was 'incomplete' and he omitted some 100 pages from the next edition (for a discussion of this case, see; Chorus 2019).

By today's standards, a direct translation that lacks quotation marks is considered plagiarism even if it contains a footnote identifying the original source.

4.3.3   Self-Plagiarism

A special case deserves our attention: using our own work without acknowledgement. This practice is known as 'self-plagiarism,' and notably is not often taken into account within the codes of conduct of most universities. Is self-plagiarism unethical? Some argue it's not. You can't, after all, 'appropriate' what is already yours. Others argue it is, as we will soon explore. At its core, it is important that readers are made aware that previously published work is recycled, even if it's the author's own work.

A differentiation must be made between 'duplication publication' (complete republication without attribution of the original), 'text recycling' (reuse of portions of one's own writings), and 'secondary publication' (republication with permission from the original publisher). An element of redundancy and wastefulness is persistent in all three forms, and in as much as they distort meta-analyses, they are all also unethical. By most accounts, text recycling and duplication publication are considered 'questionable,' though not misconduct, except when copyrights are infringed (see Habibzadeh and Winker 2009).

To showcase the sensitivities surrounding text-recycling, take the case of Peter Nijkamp, professor of economics at Amsterdam University [VU]. Throughout his career, Nijkamp was a prolific author producing at one point an astonishing output of some 50 publications a year.

In 2013, Nijkamp was accused by an anonymous whistleblower of 'excessive recycling' of his own work. A university integrity commission investigated the charge in 2014 and found that he had indeed often re-used parts of his works without proper acknowledgement. The commission condemned the practice, labeling it 'questionable research practices,' though not plagiarism (Zwemmer et al. 2015). In newspaper coverage, the scientist was accused of 'self-plagiarism'.

Nijkamp was outraged. 'Self-plagiarism is a bogus reproach', he responded. He did not see any harm in the practice of re-using one's own work. He had always acted in good faith, and had never transgressed any code of conduct, he maintained (Nijkamp 2014, p. 24). Calling the anonymous complaint a 'witch hunt', aimed at destroying his reputation, Nijkamp filed a counter charge, arguing that the complaint should not have been admissible, and that his name should be cleared (Sahadat 2015). He won the case.

While self-plagiarism may not be a transgression of a code of conduct for academic authors, this would not hold for students, who hand in their own work twice for different assignments. This is not accepted. Credits earned for assignments are given for original work only.

To check for originality, most universities utilize plagiarism detection software. This software has access to not only web publications, but also to large databases containing previous submissions, and will likely spot any similarities between two texts (see the next section for further examples).

4.4   In Other Words or in the Words of Others?

4.4.1   Stealing into Print

Many universities strongly encourage students to work together, to discuss each other's work, and give 'peer feedback.' This is aligned with the standing practice in academia to discuss unpublished work with colleagues at conferences as well as submitting manuscripts to academic journals for (anonymous) peer reviewing.

Peer reviewing is thus at the heart of academic work, but it has invited forms of plagiarism that we need to address. Submitting an unpublished manuscript or a grant proposal for review involves the risk that others will make use of it. If not by plagiarizing it, then certainly by 'stealing' valuable ideas (Lafollette 1992, p. 127ff).

Although it is unclear how often it happens, it is known to happen. In 2017, the news section of the *Annals of Internal Medicine* revealed a recent case of 'reviewer misconduct.' Michael Dansinger, the lead author of a paper that was rejected by the *Annals*, discovered that his paper appeared in a different journal a few months later, but with the names of the authors removed and replaced by the names of others. His paper had clearly been stolen by a peer reviewer of the *Annals*.

Dansinger revealed his discovery, which led to the retraction of the stolen article. But he did not want to publish the name of the plagiarist because he was not out for revenge. Instead, he wrote about the case to illustrate how and why things go wrong. Perhaps the pressure to publish was intense, Dansinger conjectured, or maybe the culture was relatively permissive such that plagiarism was not taken seriously, or maybe it was simply a matter of believing the plagiarist would not get caught. Whatever the reasons, there is an incredible risk involved in this kind of misconduct, and by revealing it, Dansinger hoped it would help deter the this kind of misconduct (Dansinger 2017, p. 143).

Community members of 'Retraction Watch,' a website dedicated to 'academic misconduct, were far less forgiving. Ralph Giorno commented: 'The repercussions need to be that ALL authors are summarily fired, then pursue libel charges. None of these people should ever practice medicine anywhere, including in a private setting.' Another commented: 'This is why review should not be anonymous' (*see Retraction Watch*, thread: 'Dear peer reviewer, you stole my paper: An author's worst nightmare', 12.12.2016).

4.4.2 Authorship

While plagiarism is unethical, even worthy of punishment, a more subtle problem lies underneath. This relates to different views of *authorship* and *text ownership*.

Typically, in Western societies, scientists are looked at from a somewhat paradoxical view. While on one hand they are considered to be autonomous authors, solely responsible for what they write, they are simultaneously expected to act as selfless parts of the 'academic community,' whose aim is the extension of our collective knowledge.

Accordingly, in the West, texts are viewed as 'private property,' while ideas are more or less considered 'common goods' (for further discussion, see; Marsh 2007). However, a nearly inverse relation exists in other parts of the world. In China, collectively accepted knowledge is associated with authorities and individuals of high esteem, and 'copying' (plagiarizing) these authors without credit can be seen as 'paying respect' to them rather than stealing (see Bloch 2007; Hsu 1981).

As a result of its recent rise as an economic superpower, China's output of scientific publications has increased immensely, but so has its number of retracted articles. Often due to plagiarism, these retractions have tainted the reputation of Chinese research. Authorities in China began to recognize this problem and are now implementing strict plagiarism policies (for a discussion, see; Gray et al. 2019).

4.4.3   Two Cases

Plagiarism software is good at spotting similarities between texts, not at identifying who plagiarized whom. Consider these two cases of suspected plagiarism reported at Utrecht University (in the Netherlands).

In the first case, the plagiarism detection software revealed the following similarities between the papers of two undergraduate students:

- Identical title
- A match of 66% in the second sentence
- Identical quotation [source correctly identified]
- A match of between 66 and 77% in the following three sentences
- 100% match in next sentence
- A match of 82% in the subsequent sentence
- A match of 90% in the main question
- 4 more sentences paraphrasing another source, with a 100% match between the two papers
- Several more sentences matching between 75 and 100%
- In the conclusion two sentences had a 70–80% match (Fig. 4.2)

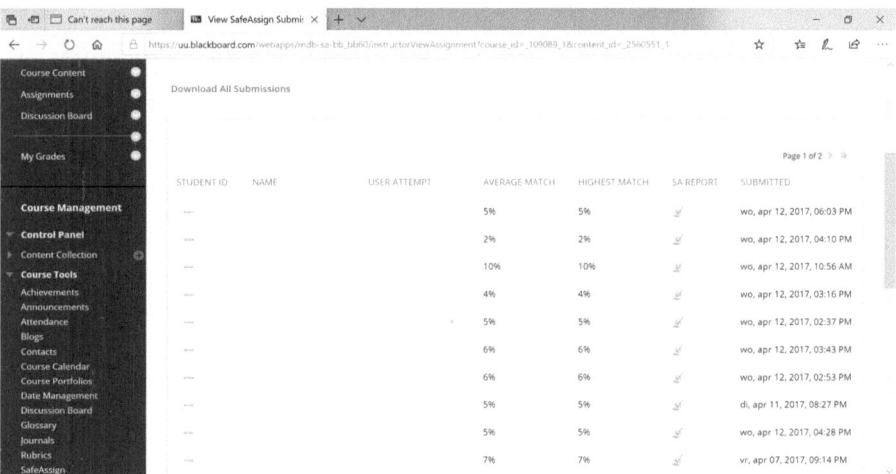

Fig. 4.2 Teacher's view of student submissions to a plagiarism detection software as presented in Blackboard. The names of the students are redacted. The two middle columns give percentages of 'matching texts'. In this sample they range between 4 and 10%, indicating very low or no occurrence of plagiarism in the submissions

In sum, the two papers had an overlap of 20%, enough to be flagged as suspected plagiarism. The case was brought before the board of examination who, following Utrecht University procedure, requested the students to respond. The first student stated that, 'I am quite startled by your request, because I am not aware of any wrongdoing. I wish to state that I have shared my paper with two of my fellow students for review, and it may be that they have copied parts of it.' The second student replied, 'I assume that my paper is coupled with that of [Student 1], who is not to be blamed. She sent her finished work to me and I may have focused a bit too much on her paper, by free riding on it.'

The second student was found guilty of plagiarism. This student received an official reprimand; her exam was annulled and she was removed from the course for the duration of a year. No action against the first student was undertaken however, even though she had actively made plagiarism possible by sending her paper to the other student, and the code of conduct of this university rules that 'inciting of plagiarism' is also punishable.

In a second, somewhat less clear-cut case, the plagiarism detection program again revealed substantial overlap between two papers, and even to a much higher degree (49%). Suspicion of misconduct was raised. Again, the students were requested to respond.

When they appeared before the board of examination, they declared that the overlap between their papers could be explained in part because they had worked in close collaboration with each other, and in part because they had shared a document on Facebook, co-authored by the two of them. Neither had plagiarized the other, they claimed.

In this case, the board ruled that as far as their close collaboration was concerned, it could *not* be considered plagiarism, but rather 'inadequate course preparation' (they were supposed to write an individual paper). As to sharing a document, it *was* ruled plagiarism because 'no adequate references to the source were given.' The students were not removed from the course but both received an official reprimand and had to rewrite the paper (Box 4.3).

4.4.4   Priority Disputes

Another form of appropriation demands our attention. It is commonly referred to as a 'priority dispute.' Typically, in such cases, one author accuses another of stealing their pre-published ideas or discoveries, claiming priority over the idea in dispute. The history of science is full of these disputes. We shall briefly examine one such example below.

In the last decades of the nineteenth century, years before he became a household name as the founder of psychoanalysis, Sigmund Freud befriended a Berlin otolaryngologist (ear, nose, and throat specialist) by the name of Wilhelm Fliess. Fliess advocated for the idea that all human beings are bisexual in nature. Furthermore, he proposed that the major events in life are 'predetermined' by two biological cycles,

Box 4.3: 'Appropriation, but not Plagiarism'

Melanie has had great difficulties finishing her BA thesis. Her teacher has already pointed out several flaws in the first version and warned her that unless she substantially changes the design, she will fail. With the deadline looming, Melanie doesn't know what to do. She considers the following strategies:

Strategy 1: She'll ask a fellow student to discuss the thesis with her and help her identify its weaknesses. Melanie will cook dinner while her friend reads the draft. During dinner they'll discuss any possible changes and revisions.

Strategy 2: She'll ask to look at the theses of two of her friends who have already finished to see what they did differently.

Strategy 3: She'll pay for the services of a professional agency. They say the work is produced by certified teachers and charge a fee of 250 Euros. They promise to rewrite her thesis in a week, based on her first draft.

Strategy 1 is entirely legitimate and should in fact be encouraged. **Strategy 2** is also legitimate, provided Melanie doesn't copy these theses. **Strategy 3** is cheating.

one lasting 23 days, the other 28. Fliess discussed these ideas with Freud but failed to publish them until much later, when their friendship had already withered away.

In 1903, psychologist Hermann Swoboda, published a book on a distinctly Fliessian notion of 'periodicity,' and a year later philosopher Otto Weininger published a highly successful book that discussed, among other things, bisexuality. Fliess recognized these as his unpublished ideas, and he suspected foul play.

In a letter dated July 20th, 1904, Fliess demanded that Freud explain himself. In a response sent on July 27th, just a week later, Freud sheepishly admitted to having discussed Fliess' ideas with Swoboda, who had subsequently passed on the information to his friend Weininger. He acknowledged Fliess' 'priority rights' with respect to bisexuality, but he denied that he had given any *detailed* information to Swoboda (Masson 1985, pp. 462–8). In other words, both Swoboda and Weininger had merely acted upon a hint and worked out the rest on their own. This did not convince Fliess, who published an angry pamphlet accusing Freud of betrayal for giving away vital information that belonged to him.

Had Freud been careless with his former friend's ideas? Was he responsible for the 'theft' (or plagiarism) of Fliess' intellectual property? In a paper discussing the case in detail, Michael Schröter (2003) compared the content of both Swoboda and Weininger's books and found that they both contain ideas resembling Fliess' work. That being said, Schröter noted that Swoboda's work contained notable differences as well, and Weininger's 'appropriation' of Fliess' ideas were few and far between, effectively as little as a mere sentence. However, Schröter does admit that Fliess' accusations, although exaggerated, do contain 'a kernel of truth.'

Fig. 4.3 Freud (left) and
Fliess (right), here still
friends, ca. 1895. (Source:
Ernst Freud 2006, p. 156)

Fliess' priority claim (right of intellectual property) stands as an example of how the majority of priority disputes occur. Seldom do we see the theft of a final product. More often, authors find themselves grappling with similar ideas at the same time as one another, and subsequently accusing the other of stealing their ideas. Historians argue that, in these types of cases, the ideas were essentially up 'in the air.' Does that mean priority disputes aren't 'real' disputes? No. As with other forms of plagiarism, priority disputes point to the fact that ideas cannot develop in isolation; they need *other* ideas to develop, and often competition plays a major role in these developments (for further discussion, see Michael White's 2002 book *Rivals*). Conscientious authors give credit where credit is due – not merely out of politeness or to avoid unpleasant priority disputes, but precisely because acknowledging contributions allows them to establish their *own* claims (Fig. 4.3, Box 4.4).

4.5   When Do Intentions Come into Play?

4.5.1   Intentions Matter

Not everyone who commits the 'unoriginal sin' of plagiarism does so on purpose. Some don't fully grasp that what they're doing is wrong and when caught plagiarizing, their defense often reads like this: I may have been sloppy, but I did not *intend* to plagiarize. Well then, let's explore a few examples (drawn from Kolfschooten 2012) to see whether that defense holds any water.

Box 4.4: 'Plagiarism, but not Intentional'
Oddly enough, Freud almost committed plagiarism himself with respect to Fliess' ideas on bisexuality. It happened during a discussion when they were still friends. Freud explained to Fliess that he believed that the problem of neurosis could be resolved if the individual's bisexuality were taken into account. Fliess allegedly responded matter-of-factly: 'That's what I told you two and a half years ago, but you would have none of it then' (Freud 1901/1960, p. 144).

Have you ever had a seemingly novel idea or solution to a problem suddenly come to mind? Of course. Have you ever then remembered, or been informed, that the idea or solution was in fact something you had previously heard? The answer is likely yes. The act of forgetting something that you've heard and the subsequent reappearance of that memory believed to be your own idea is known as *cryptomnesia*, and it may cause *inadvertent plagiarism*.

In a series of psychological experiments, researchers were able to produce cryptomnesia in a group of students. The participants were tasked to take turns spontaneously generating lists of items in specific categories such as 'sports' and 'animals.' When asked to recall the items they had listed before adding new ones, subjects would 'appropriate' items others had produced earlier during the session, and presented them as their own (see; Brown and Murphy 1989).

4.5.2 Plain Sloppiness or Culpable Carelessness?

Our first example is focused on a senior lecturer that contributed several chapters to a handbook that was flagged for possible plagiarism. The lecturer maintained before the integrity commission that he had relied on the help of a student who, it turned out, had copied parts of a text verbatim, without reference. When questioned, the student admitted to having been 'sloppy.' Was it then the student's fault? The commission determined that because the chapters were published under the lecturer's name, the lecturer was responsible, not the student. The author was found guilty of plagiarism because they should have checked the student's work. Ultimately, the case was ruled 'nonintentional' and therefore held less culpability.

The next example followed a junior researcher who submitted a paper to a conference. A reviewer detected 'significant similarities' between the junior researcher's paper and a paper published by a senior researcher the year before. After an investigation, it was determined that the junior researcher had indeed copied portions of the other's work. Despite asking the senior researcher for permission to use certain tables, the junior researcher failed to identify the source properly. The junior researcher's contract was subsequently terminated on the grounds of 'unsuitability, in casu plagiarism.' The senior researcher thought it wasn't a case of willful plagiarism, but a case of 'bad citation by a rookie.'

	Intention to commit plagiarism	No intention to commit plagiarism
Severe infringement of university policy	Stealing source material	Faulty/sloppy referencing Inadvertent plagiarism
Light infringement of university policy	Patch-writing Improper paraphrasing	Minor transgressions and irregularities, sloppy referencing

Fig. 4.4 Taxonomy of Plagiarism (after Walker 1998)

Box 4.5: 'List of Objects That Can Be Plagiarized'
- Ideas;
- Research findings;
- Phrases;
- Entire texts;
- Charts;
- Illustrations (including photographs, scans, and figures);
- Lecture notes and PowerPoint slides;
- Lecture summaries;
- Exams.

Note that lecture summaries, PowerPoint slides, and exams, all content university students regularly encounter, are specifically included in this list. Students have the right to inspect exams and use slides and summaries as reference material, but to plagiarize their content is considered misconduct in most universities.

Notably *not* on this list are particular expressions which are often too well-known to be considered plagiarizable – for example expressions like 'resilience' or 'unconscious.' However, if these expressions relate to specific authors, they would need referencing. For example: 'paradigm shift' (Kuhn), 'survival of the fittest' (Darwin), or 'unintended consequences' (Merton).

Additionally, 'reference lists' are not considered plagiarizable because typically such lists are supposed to act as a safeguard *against* plagiarism. Thus, when a classroom of students work on an assignment comparing several articles, their bibliographies will be (nearly) identical. Plagiarism detection software will still spot these 'similarities,' but a teacher will recognize that they aren't 'matches' for plagiarism. It's a different story, though, if two students work on an open-ended assignment and their bibliographies come back identical; then suspicion may arise.

The final example followed 'an Egyptian serial plagiarizer.' In 2003, a mathematician learned through a colleague that one of his papers had been plagiarized by a mathematician in Egypt. When the journal was alerted, the fraudulent paper was quickly withdrawn. Further investigation into the Egyptian mathematician exposed routine plagiarism. Their articles were withdrawn and they were put on the publication blacklist (Fig. 4.4, Box 4.5).

4.6   Factors That Facilitate Plagiarism

4.6.1   Factors

The cases discussed in the previous sections reveal how different intentions play a role in plagiarism (and how these intentions are weighed into establishing a verdict), but also how specific scientific customs and traditions (such as the difference between US and Chinese citation practices) must be taken into account. In short, a variety of factors contribute to the occurrence of plagiarism and awareness is key. In this section, we briefly review relevant literature that points to the most important factors currently inciting students to commit plagiarism (or inversely keep them away from it).

4.6.2   Experience

A number of authors writing on the subject perceive there to be a higher prevalence of plagiarism among junior (first and second year) undergraduate students. They attribute this to a lack of experience in academic writing (Park 2003). The underlying idea being that senior (third and fourth year) undergraduate students are supposed to have developed better writing skills, and should therefore be better able to avoid plagiarism. Walker (2010), however, found just the opposite. Students 20 years and younger plagiarized significantly *less* than students 21–30 years of age. Walker reasoned this was likely due to feeling more pressure to perform.

Regardless of whether or not junior students plagiarized more than senior students, the lack of experience may still elicit feelings of uncertainty surrounding issues of plagiarism. This uncertainty was expressed by a UK master's student, who was interviewed on the prevalence of cheating and plagiarism, noting: 'You don't know what is cheating, if you've got an idea of an article, or if it is your own idea and you write it down in your own words' (Ashworth et al. 1997, p. 191).

4.6.3   Externalization

Lack of academic experience and feelings of powerlessness and uncertainty may well contribute to an 'externalizing' of the problem. As some students expressed, plagiarism is something university professors find important, but isn't something perceived to be 'a real problem.' Often, students consider plagiarism 'a minor offence' at best, even 'no big deal' (Park 2003). The concept of intellectual property is not (yet) common knowledge to these students; and as Power (2009) concluded following interviews with students on their perceptions of plagiarism, it is rather 'imposed on them by authorities or other people in power outside of themselves' (p. 654).

4.6.4 Pressure to Perform

Most, if not all universities across Europe and the US have become very demanding, competitive institutions in the last few decades. Universities have begun nurturing a 'culture of excellence' that requires students to hand in an abundancy of assignments over short periods of time, further expecting a continuous performance at the highest level. As identified by Comas-Forgas and Sueda-Negre (2010), the three most relevant causes of plagiarism are: lack of time to carry out academic assignments, poor time-management and personal organization, and performance pressure.

4.6.5 Availability

The internet allows access to a seemingly infinite number of sources. Copy-pasting from internet sources (*cyber-plagiarism*) has become almost too easy – were it not for plagiarism detection software. Inversely, it is becoming difficult to *not* rely on internet sources, and considerable effort gets put into hiding that fact. As one student put it: 'Things are a lot easier to get away with on the internet if you wanted to (give false information for example). But copying work without sourcing it is easier from books in my opinion, as universities have methods of screening essays for plagiarized works through the internet' (Selwyn 2008, p. 473). The lesson to be drawn from this is that acquiring good writing skills should be a priority, since poor writing has become both easy and dangerous.

4.6.6 Faulty Teachers

Some students blame their teachers, or the educational system as a whole, for failing to properly define the difference between legitimate use of source material and plagiarism: '[Teachers] just say "Don't plagiarize." But they never tell you what to do to not plagiarize' (Power 2009, p. 655). There may be more to this than a simple justification of laziness, as another student explains: 'Well, listen, I'm terribly confused what it actually means – I mean that might sound stupid: there's a policy that... the wholesale copying is obviously quite obvious, but there's a hell of a lot of grey area in between that I really don't even understand' (Gullifer and Tyson 2010, p. 470).

4.6.7 Cultural Expectations

In Sect. 4.3 of this chapter, we touched upon the different cultural perceptions of authorship, particularly between Western (in our case, the US and Europe) and Eastern (in our case, China) societies. 25 years ago, Deckert (1993) found that

Chinese students enrolled at an English college in Hong Kong hardly had any notion of what plagiarism was at all, and held a very a different view on authorship compared to American students. Walker (2010) confirmed what others have since observed, namely that international students at Western universities show higher rates of plagiarism (and more serious forms as well).

Of course, these (isolated) findings should not be taken as incitements of cultures in general, but rather of different norms and expectations with regard to publication behavior and source refencing. Indeed, Hu and Lei (2015) acknowledge that Chinese student often have limited knowledge of Anglo-American intertextual conventions, and emphasize the need to develop effective instructional strategies to master these literary practices, and help 'raise international students' awareness of cross-cultural differences in intertextuality. Further, the need to equip them with the requisite skills and strategies to engage in legitimate textual appropriation' (2015, p. 255) (Box 4.6).

Box 4.6: 'Who's to Blame? A Dilemma'

Jack and Jill are third year sociology students. They have been working independently on an assignment about the prisoner's dilemma game. When the deadline comes, Jack is the first to upload his paper, followed by Jill a few minutes later.

While Jack and Jill put a great deal of effort into their papers, the plagiarism software detects a 58% match – meaning there are many examples of nearly identical sentences, and the two papers even have a matching structure. It appears obvious that one has plagiarized the other. The question is: who did the copying?

When asked for clarification, both Jack and Jill claimed priority and accused the other of being the plagiarizer. The teacher knows the two students well and is aware that Jack is a rather weak student. The paper appears almost 'too good' for Jack, and seems much more at Jill's level. But the teacher can't prove this.

What should the teacher do? Of the three options presented below, which do you think is the most fair? Discuss your choice with your classmates.

1. It may be likely that Jack and Jill have struck a deal to blame each other to ensure the real offender goes unpunished. With no available proof of who plagiarized whom, the teacher must accept the situation as it is and grade both papers.
2. There is undeniable proof that one student has copied from the other. Although it cannot be established who is the plagiarizer, any participation in plagiarism is considered misconduct and the teacher determines both students are in the wrong and as such, both must do the assignment again.
3. One of the papers is undeniably copied from the other. It's much more likely that Jack has plagiarized Jill. Therefore, Jack must do the assignment again, with Jill is also reprimanded since she has allowed her paper to be plagiarized. She gets a deduction in her grade. Instead of receiving an A+, as she was initially assessed, the paper gets a B−.

4.7   Conclusions

4.7.1   Summary

In this chapter we have explored the litany of problems surrounding plagiarism, which we understand to be the *appropriation of someone else's work (or ideas) without appropriate referencing* of the original sources.

We have learned to distinguish plagiarism from legitimate forms of referencing, including *quotations*, *paraphrasing,* and *summarizing*. There are ample *publication manuals* that contain detailed rules on how to properly reference your sources, be it written or oral communications, translations, and parts of or even whole sentences. The underlying principle to remember is that your reader must be able to identify the source of every piece of information you use.

While plagiarism is morally and often legally wrong, certain cultural or social circumstances allow for a nuanced view. Thus, different views on *authorship* and especially the role of *intentions* complicate matters to the point that cases of plagiarism must be judged on an individual basis, to establish what went wrong, who is to be blamed, and why. *Cryptomnesia*, a form of inadvertent plagiarism, serves as an interesting case in point here.

Of special interest are the forms of appropriation that involve stitching together a somewhat loose patchwork of source material called *patch writing*. Patch writing borders on plagiarism because it borders on copy-paste techniques. Patch writing is considered a 'grey area' (neither wrong, nor right), although it involves stylistic matters too.

Furthermore, we discussed *priority disputes* surrounding plagiarism and considered whether or not *self-plagiarism* can be regarded as bona fide plagiarism, or if it is actually a form of 'passive self-citation.'

Finally, we explored the *factors that facilitate* plagiarism, including experience with writing (or lack thereof), externalization of the problem (perceiving plagiarism to be someone else's problem), performance pressure, availability of sources (books vs. online), inadequate teachers, and cultural beliefs.

4.7.2   Discussion

The object of this chapter is to familiarize you with the problems surrounding plagiarism. By becoming aware of this issue's many facets, we hope you will be able to act responsibly and avoid being caught in a situation like the ones we explored. Plagiarism can and must be avoided, but it takes training. Everyone has a responsibility here, including you and your university.

Institutions and universities must, at a minimum, provide clear guidelines regarding their referencing requirements, and stipulate the consequences of noncompliance. Hopefully they'll do even more, and allow you and your fellow students time to adequately train yourselves. Newton et al. (2014) found that even short-duration plagiarism training programs significantly enhance students' in-text referencing skills. Offerings such as this can allow you to better understand how to properly utilize the work of others and to feel better equipped to carve out your own niche in the scientific community.

There is no better time to start thinking through these issues than now. Begin by discussing some of the issues outlined in this chapter, in particular involving the problem of authorship. Who 'owns' texts and ideas and why is that? How private is your work? What steps must you take to become the 'author' of a text?

Case Study: René Diekstra

René Diekstra, a professor of clinical psychology at Leiden University in the Netherlands, gained notoriety in the summer of 1996 after he was accused of plagiarism, and subsequently resigned.

The situation, described in detail by Frank van Kolfschooten (2012), meant the end of a highly successful academic career. Internationally recognized as a leading expert in suicidal behaviors, Diekstra was the founder of the International Academy of Suicide Research. He was also the founder of the scientific journal *Archives of Suicide Research*. He had been the manager of the 'Psychosocial and Behavioral Aspects of Health and Development' of the World Health Organization (WHO) program and was one of the first recipients of the Stengel Award, the world's most prestigious honor in the field of suicide research.

When Diekstra lost his university position in 1996, he was ousted from his organizational positions and at least one of his awards was revoked. Effectively, he became a persona non grata. How did this fall from grace come to pass?

Being a prolific writer, Diekstra had written numerous scientific publications, as well as multiple popular science and self-help books. The publication that sparked his downfall was 'No Stone Unturned' (*De Onderste Steen Boven*, 1996), written for a large non-academic public. The charge of plagiarism had been levelled by two journalists of a local Dutch weekly newspaper. Allegedly, Diekstra had copied entire sections from *How to Deal with Depression* by Bloomfield and McWilliams (1994). The journalists responded with indignation: 'Who is this Diekstra? And this man calls himself a professor!' (quoted in Danhof en Verhey 1996) (Fig. 4.5).

Despite being on vacation, Diekstra quickly responded. He admitted to having copied roughly twenty pages from *How to Deal with Depression*, but he also placed some of the blame on the publisher, suggesting a possible lapse in communication. The scandal did not end there, though. When he returned from his vacation, a subsequent publication in the same journal revealed further indications of plagiarism. Apparently, Diekstra had translated 26 pages from an unpublished manuscript by Gary McEnery and incorporated them into his self-help book 'When Life Hurts' (*Als het Leven Pijn Doet*), published in 1990.

While Diekstra was out on sick leave, an independent commission began investigating the case, and in that time, another allegation of plagiarism came to the

Fig. 4.5 Media coverage of the Diekstra case. (Source: *Leidsch Dagblad*, August 29, 1996)

forefront. This time in a scientific publication, co-authored with a colleague. When
called to testify, Diekstra admitted 'carefree' use of sources in 'No Stone Unturned'.
He denied, however, that he had plagiarized McEnery, who was credited as a co-
author on the title page of 'When Life Hurts' (though the cover of the book showcased only Diekstra's name). For the charge of plagiarism in the scientific article, he
flatly denied having any knowledge thereof and entirely blamed his co-author.

By December of 1996, the commission concluded that the allegations of plagiarism in Diekstra's self-help books 'When Life Hurts' and 'No Stone Unturned'
were legitimate. Regarding the scientific article, the commission was convinced
Diekstra was at least *partly* responsible and that his position as professor at Leiden
University was compromised (Hofstee and Drupsteen 1996). As a result, Diekstra
handed in his notice, though he disagreed that he had committed 'scientific misconduct.' Rather, he felt he had been 'sacrificed.'

Several of Diekstra's colleagues at Leiden University came to his defense. They
published a 272 page plea, arguing that the executive board of the university should

reconsider the commission's verdict. They stated that the report of the commission was sloppy and the punishment (dismissal) 'disproportional' (Dijkhuis et al. 1997). The following year, in 1998, Diekstra published an autobiographical account of the affair titled 'O Holland, Land of Humiliation!' (*O Nederland, Vernederland!*), which detailed his feelings of being wronged. He notes how the investigative commission had acted in 'bad faith' and 'twisted the facts,' but that the damage had been done: he was ridiculed in public by his friends and his former colleagues turned their backs on him. Throughout it all, Diekstra maintained his innocence, and that the allegations of plagiarism were false (though he again admitted 'carefree handling' of other people's materials).

His self-defense strategy ran along the following three lines: (1) popular science books are exempt (at least to a degree) from the same standards as scientific publications; (2) the plagiarized parties had been offered financial compensation; (3) he had not *intended* to steal from others. Therefore, he argued, the accusation of plagiarism did not stand, and was at best simply 'sloppiness.'

None of these arguments, nor his colleagues plea, were accepted, and the conviction remained. Diekstra, however, continued to seek rehabilitation for many years. By 2003 he was still attempting to sue Leiden University, but to no avail.

In the years following 1997, Diekstra became the Director of the Center for Youth and Development in The Hague, the Netherlands and between 2004 and 2011, was head of the Social Science Department and professor of psychology at the Roosevelt Academy in Middelburg. However, as he was reassembling his life, Diekstra's past continued to haunt him. In 2004, his position as a professor at Roosevelt caused a stir because according to the responsible parties, his reappointed as a professor was unlawful.

Willem Koops, the acting Dean of the Faculty of Social Sciences at Utrecht University during this time, strongly opposed Diekstra's appointment at Roosevelt. He was quoted as saying: 'Diekstra views himself as the victim instead of the offender (…) That's why I think that Diekstra cannot be engaged as role model in the training of students and PhDs' (Kolfschoten 2012, p. 78).

Assignment

1. Examine Diekstra's three self-defense arguments. Do any of his arguments exonerate him (partially or fully) from the accusation of plagiarism? Does it make a difference when a clear definition of plagiarism is lacking (as was the case in 1996)?

2. Diekstra felt that he was being 'persecuted' and treated 'unfairly.' He complained that the Dutch people were unwilling to forgive him, and that they would never let the past go. To what degree (or for how long) should misconduct be considered a stain on one's reputation? Is there a point at which someone found guilty of misconduct should be allowed to start with a clean slate? What conditions should this depend upon?

3. If someone accused of plagiarism were to ask you for advise, how would you have respond? Would you recommend a different defense strategy?

Suggested Reading

A very good introduction into debates surrounding plagiarism is offered by R.M. Howard, *Standing in the Shadow of Giants: Plagiarists, Authors, Collaborators* (1999). Another notable contributions comes from M.C.Lafollette, *Stealing into Print. Fraud, Plagiarism, and Misconduct in Scientific* (1992). Finally, we recommend Ashworth et al. 'Guilty in Whose Eyes? University student's perception of cheating and plagiarism in academic work and assessment' (1997) for an excellent overview of student perspectives on plagiarism.

References

American Psychological Association. (2010). *Publication manual of the American psychological association* (6th ed.). Washington, DC: American Psychological Association.

American Sociological Association. (2014). *Style Guide*, 5th ed. Washington.

Ashworth, P., Bannister, P., & Thorne, P. (1997). Guilty in whose eyes? University student's perception of cheating and plagiarism in academic work and assessment. *Studies in Higher Education, 22*(2), 187–203. https://doi.org/10.1080/03075079712331381034.

Bloch, J. (2007). Plagiarism across cultures: Is there a difference? *Indonesian Journal of English Language Teaching, 3*(2), 1–13. https://doi.org/10.25170/ijelt.v3i2.133.

Brown, A. S., & Murphy, D. R. (1989). Delineating inadvertent plagiarism. *Journal of Experimental Psychology: Learning, Memory, and Cognition, 15*(3), 432–442.

Chorus, R. (2019). *Alphons Chorus: beeld van de mens en psycholoog.* Amsterdam: SWP Publishing.

Comas-Forgas, R., & Sureda-Negre, J. (2010). Academic plagiarism: Explanatory factors from students' perspective. *Journal of Academic Ethic, 8*(3), 217–232. https://doi.org/10.1007/s10805-010-9121-0.

Dansinger, M. (2017). Dear plagiarist: A letter to a peer reviewer who stole and published our manuscript as his own. *Annals of Internal Medicine, 166*(2), 143. https://doi.org/10.7326/M16-2551.

Deckert, G. D. (1993). Perspectives on plagiarism from ESL students in Hong Kong. *Journal of Second Language Writing, 2*(2), 131–148. https://doi.org/10.1016/1060-3743(93)90014-T.

Freud, E. (Ed.). (2006). *Sigmund Freud: Sein Leben in Bilderns und Texten.* Frankfurt am Main: Fischer.

Freud, S. (1901/1960). The psychopathology of everyday life. In J. Strachey (Ed.), *The standard edition of the complete psychological works of Sigmund Freud* (Vol. 6). London: Hogarth Press.

Gray, G. C., Borkenhagen, L. K., Sung, N. S., & Tang, S. (2019). A primer on plagiarism: Resources for educators in China. *Change: The Magazine of Higher Learning, 51*(2), 55–62. https://doi.org/10.1080/00091383.2019.1569974.

Gullifer, J., & Tyson, G. A. (2010). Exploring university student's perceptions of plagiarism: A focus group discussion. *Studies in Higher Education, 35*(4), 463–481. https://doi.org/10.1080/03075070903096508.

Habibzadeh, F. A., & Winker, M. (2009). Duplicate publication and plagiarism: Causes and cures. *Notfall + Rettungsmedizin, 12*(415), 415–418. https://doi.org/10.1007/s10049-009-1229-7.

Howard, R. M. (1999). *Standing in the shadow of giants: Plagiarists, authors, collaborators.* Stamford: Ablex Publishing Company.

Hsu, L. K. (1981). *Americans and Chinese.* Honolulu: University Press of Honolulu.

Hu, G., & Lei, J. (2015). Chinese university students' perceptions of plagiarism. *Ethics & Behavior, 25*(3), 233–255. https://doi.org/10.1080/10508422.2014.923313.

Kolfschooten, F. (2012). *Ontspoorde wetenschap* [Science derailed]. Amsterdam: De Kring.

Lafollette, M. C. (1992). Stealing into print. In *Fraud, plagiarism, and misconduct in scientific publishing.* Berkeley: University of California Press.

Lahusen, B. (2006). Goldene Zeiten: Anmerkungen zu Hans-Peter Schwintowski, Juristische Methodenlehre, UTB Basics Recht und Wirtschaft 2005. *Kritische Justiz, 39*(4), 398–417. https://www.jstor.org/stable/26425890.

Marsh, B. (2007). *Plagiarism. Alchemy and remedy in higher education.* New York: State University of New York Press.

Masson, J. M. (1985). *The complete letters of Sigmund Freud to Wilhelm Fliess 1887–1904.* Cambridge: Harvard University Press.

Newton, F. J., Wright, J. D., & Newton, J. D. (2014). Skills training to help avoid inadvertent plagiarism: Results from a randomized control study. *Higher Education Research and Development, 3*(6), 1180–1193. https://doi.org/10.1080/07294360.2014.911257.

Nijkamp, P. (2014). Het is van den zotte dit zelfplagiaat te noemen [It's absurd to call this self-plagiarism]. *Advalvas, 10*, 23–25.

Park, C. (2003). In other (people's) words: Plagiarism by university students – Literature and lessons. *Assessment & Evaluation in Higher Education, 28*(5), 471–488. https://doi.org/10.1080/02602930301677.

Power, L. G. (2009). University's students' perception of plagiarism. *Journal of Higher Education, 80*(6), 643–662. https://doi.org/10.1080/00221546.2009.11779038.

Rogoff, B. (2003). *The cultural nature of human development.* Oxford: Oxford University Press.

Sahadat, I. (2015, March 21). 'Natuurlijk knip ik uit eigen werk' [Sure I copy-paste my own work]. Interview with Peter Nijkamp, *De Volkskrant.*

Saunders, J. (2010). Plagiarism and the law. *Learned Publishing, 23*(4), 227–292. https://doi.org/10.1087/20100402.

Schröter, M. (2003). Fliess versus Weininger, Swoboda and Freud: The plagiarism conflict of 1906 assessed in the light of the documents. *Psychoanalysis and History, 5*(2), 147–173.

Selwyn, N. (2008). 'Not necessarily a bad thing…': A study of online plagiarism amongst undergraduate students. *Assessment & Evaluation in Higher Education, 33*(5), 465–479. https://doi.org/10.1080/02602930701563104.

Walker, J. (1998). Student plagiarism in universities: What are we doing about it? *Assessment & Evaluation in Higher Education, 17*(1), 89–106. https://doi.org/10.1080/0729436980170105.

Walker, J. (2010). Measuring plagiarism: Researching what students do, not what they say they do. *Studies in Higher Education, 35*(1), 41–59. https://doi.org/10.1080/03075070902912994.

White, M. (2002). *Rivals. Conflict as the fuel of science.* London: Vintage Books.

Wootton, D. (2015). *The invention of science. A new history of the scientific revolution.* London: Penguin.

Zwemmer, J., Gunning, J.W. & Grobbee, R. (2015, February). *Report concerning references cited in the work of Professor P. Nijkamp* (Unpublished findings of the integrity committee, Amsterdam).

References for Case Study: René Diekstra

Bloomfield, H. H., & McWilliams, P. (1994). *How to deal with depression*. Los Angeles: Prelude Press.

Danhof, E., & Verhey, E. (1996, August 24). René Diekstra's plagiaat: tienduizend dollars voor dertien overgeschreven pagina's [René Diekstra's plagiarism: Ten thousand dollars for thirteen copied pages]. *Vrij Nederland, 20*.

Diekstra, R. (1990). *Als het leven pijn doet* [When life hurts]. Utrecht: Bruna.

Diekstra, R. (1996). *De onderste steen boven* [No stone unturned]. Utrecht: Bruna.

Diekstra, R. (1998). *O Nederland, vernederland* [O Holland, land of humiliation]. Utrecht: Bruna.

Dijkhuis, J., Heuves, W., Hofstede, M., Janssen, M., & Rörsch, A. (1997). *Leiden in last. De zaak Diekstra nader bekeken* [Leiden in distress. A closer look at the Diekstra case]. Leiden: Elmar.

Hofstee, W. K. B., & Drupsteen, Th. G. (1996). *Rapport van de Onderzoekscommissie inzake beschuldigingen Diekstra* [Report on the commission investigating the allegations against Diekstra]. Leiden: University of Leiden.

Van Kolfschooten, F. (2012). *Ontspoorde wetenschap* [Science derailed]. Amsterdam: De Kring.

Chapter 5
Fabrication and Cheating

Contents

Electronic Supplementary Material: The online version of this chapter (https://doi.org/10.1007/978-3-030-48415-6_5) contains supplementary material, which is available to authorized users.

© The Author(s) 2020
J. Bos, *Research Ethics for Students in the Social Sciences*,
https://doi.org/10.1007/978-3-030-48415-6_5

After Reading This Chapter, You Will:

- Know what fabrication is
- Be able to distinguish between fabrication and other forms of fraud
- Understand how fabrication impacts the social sciences
- Comprehend how institutions respond to cheating and fabrication

Keywords Canvassing · Cheating · Cheating crisis · Crisis of confidence · Deception · Fabrication · Factor loading · Fake science · Forgery · Fraud triangle · Free riding · Ghostwriting · Hoaxing · Honor code · Linguistic analysis · Paper mill · Proctoring · Publication pressure · Whistleblowing

5.1   Introduction

5.1.1   The Expert's Sin

If plagiarism is the crime of the novice, then fabrication is the expert's specialty. Unlike 'falsifying' (the subject of the next chapter), which is the deliberate *mis*representation of data, fabrication involves the presentation and reporting of fake or non-existent research procedures, data, and findings. Fabrication is a form of cheating. It is about turning science upside down, it starts, rather than ends, with the answer to a question.

Fabrication probably occurs less frequently than plagiarism, but it is a much more serious form of misconduct. In a systematic review of the literature on the prevalence of scientific misconduct, Daniele Fanelli (2009) found that 2% of scientists admit to serious forms of misconduct, such as fabrication or modifying data, at least once. Additionally, 14% of respondents observed this misconduct in colleagues. The discrepancy between these findings, in which people perceive themselves to be more honest than their peers, is known as the 'better than average effect' (see; Festinger 1954).

There is another bias in these figure. Self-reporting tends to underestimate the real frequency of scientific misconduct. The incidence of fraud may be higher than we know. This triggers one's imagination, spawning a number of questions: How many more cases actually exist that just haven't been discovered yet? How likely is it that fabrication eventually comes to be discovered (unlike with plagiarism, there is no 'fabrication detection software' available)? How vulnerable are the social sciences to this type of fraud? Or is it more inherent to certain research environments? As Jennifer Crocker (2011) notes, fraud 'starts with a single step,' an observation relevant to the infamous case of Diederick Stapel, which we will discuss in our case study.

While the take-home message of this chapter is that we must arm ourselves against these forms of fraud, we should also realize that the dividing line between proper and fraudulent behavior can be thin. Many fraudsters start their criminal careers with small transgressions that gradually increase in scale, especially when no one stops them in their tracks.

In this chapter, we will explore three specific forms of fabrication: *forgery, cheating,* and *ghostwriting*. We will then discuss the factors that facilitate fabrication, concluding with an examination of institutional counterstrategies.

5.2 Forgery

5.2.1 The Manufacturing of Science

The invention of complete datasets, and the fabrication of entire cohorts of respondents and their responses may be more difficult to accomplish than appears at first sight. A 'successful fraud' must not only know what 'good results' are but must also know how data convincingly corroborates conclusions. How does forgery work? What are its tell-tale signs? And what happens once the fraud is exposed?

5.2.2 Telltale Signs of Fraud

Diederik Stapel, a prolific writer and charismatic figure in social psychology in the Netherlands, succeeded in conning many of his colleagues with what is considered one of the greatest cases of fraud in the social sciences. He was exposed after three junior colleagues found his findings suspicious. The affair created a shockwave throughout the world of social psychology, leading to what is called 'a crisis of confidence' with the public.

Were there any tell-tale signs in Stapel's publications that indicated fraud? A commission that later investigated his work found sloppy mistakes and 'unbelievably high factor loadings' (a statistical term understood as an indication of an item's relative importance).

The question was raised as to why peer reviewers had never noticed his fraud. Interestingly, the tell-tale signs of fraud were revealed in a *linguistic* analysis of his work, in which Stapel's fraudulent studies were compared with his genuine work. The fraudulent writing contained 'significantly higher rates of terms related to scientific methods and empirical investigation,' suggesting that fraudulent papers involve an 'overproduction of scientific discourse' (Markowitz and Hancock 2014, p. 2). In other words, Stapel's studies were not only *too good to be true*, they were often also *too wrapped up in scientific jargon to be true*.

5.2.3   The Student Who Almost Got Away with It (Until Another Student Blew It)

The following story, reported by Jesse Singal in *New York Magazine* on May 29th 2015, gives us insight into the case of a student involved in data fabrication, and of another student who blew the lid off.

Michael LaCour was a political science student at UCLA (University of California, Los Angeles), who rose to fame in 2013 when he discovered information that contradicted everything that was then known about 'canvassing.' 'Canvasses' are short conversations between people, with one person attempting to persuade the other, often occurring during political campaigns. Typically, these forms of contact are known to have little to no lasting effect on an individual's political ideals. That is, until LaCour claimed to have found that brief talks (lasting roughly 10 min) about marriage equality, with a canvasser who revealed during the chat that they are gay, had a significant, lasting effect on the voter's views (as measured by an online survey administered before and after the conversation).

LaCour managed to get his results published in the prestigious journal *Science* (with senior co-author Donald Green). It instantly attracted nationwide attention. When LaCour discussed his work with David Broockman, a third-year political science grad student at Berkeley, the latter was so impressed that he sought to replicate the study. It wasn't long before Broockman became suspicious. Not only did he fail to replicate the original findings, he also found irregularities in LaCour's original data. They were 'too orderly.' When he subsequently contacted the firm that supposedly performed the surveys for LaCour, he learned that they had undertaken no such survey.

Broockman discussed his misgivings with Neil Malhotra, professor at Stanford's business school, who advised him not to blow the whistle to avoid possible repercussions. Broockman decided to come forth with his findings regardless, contacting LaCour's co-author Green. Green confronted LaCour, who failed to alleviate any of his doubts. Thereupon Green requested that their paper be retracted (against LaCour's wishes).

The story ended badly for LaCour. An offer to become an assistant professor at Princeton was rescinded. But Green too suffered repercussions, seeing a fellowship worth $200,000 fall through. Broockman, on the other hand, got a tenured-track professorship at Stanford University.

With the event behind him, Broockman spoke with Jesse Singal, the journalist who covered the case, reflecting on his experience as the whistleblower. Broockman compared it to what he went through when, as a teenager, he came out as a gay. 'Part of the message that I want to send to potential disclosers of the future is that you have a duty to come out about this, you'll be rewarded if you do so in a responsible way […].' (quoted in Singal 2015).

5.2.4 Whistleblowing

Whistleblowers such as Broockman fulfill an important but often risky and thank-less role in science. Unlike in Broockman's case, the outcome for many whistle-blowers falls far short of a happy ending. Perhaps this is because in many cases, whistleblowers are in a vulnerable position (which was why Malhotra advised Broockman against it).

Consider the case of Saskia Vorstenbosch, a PhD student at the Leiden University Medical Centre (LUMC) in the Netherlands (the following details are drawn from reporting by De Vrieze, 2017).

Vorstenbosch worked with a cell biologist in the early 2000s. One day, while preparing a presentation, she discovered anomalies in a number of experiments per-formed by the biologist. She 'started digging' and found evidence that suggested some of the data had been 'manipulated.' After reexamining her findings, Vorstenbosch reported her suspicions to the head of the department, who was reluc-tant to start an investigation. The department head only initiated an investigation after Vorstenbosch insisted she would report the case with or without him.

After an 18-month investigation, the integrity commission at LUMC indeed found irregularities, but only in one of the biologist's papers. Dissatisfied with this outcome, Vorstenbosch took the report to the integrity commission of the Dutch National Academy of Sciences (KNAW), who researched the case more thoroughly and concluded that other forms of misconduct had taken place. The commission advised that four of the biologist's articles be retracted. By this point, the researcher no longer worked in the Netherlands and managed to keep her name out of the press. No actions were taken against her, nor were any more of her publications withdrawn.

Tragically, Vorstenbosch, whose area of study was partly based on the biologist's research, had only achieved in undermining her own work, because her data were now also contaminated. She withdrew from science altogether even though LUMC offered her a new PhD trajectory. Speaking with a reporter about her experience, she reflected: 'People don't seem eager to undertake action [against fraud] because it might damage their own name. It's true that I too have been damaged, but should that have been reason for me to say: I'll leave it like that, I'll just keep silent? Sure enough [after fraud is discovered] publications are going to be withdrawn, but you don't want your name attached to something which you know is not true, do you?' (Fig. 5.1)

Fig. 5.1 *The Whistleblower*

5.2.5 Exposing Fraud

From cases such as these, two provisional conclusions can be drawn. The first is that scientific frauds oftentimes betray themselves, leaving traces of their misdeeds. That's because fraudulent researchers mimic real research. However, since they work backward, from conclusions to data, their results are often unnaturally orderly. Ironically then, a successful fraud must build in imperfections and create small deviations from the expected outcome. This might actually involve more work than performing real research.

The second conclusion is that *exposing* fraud can prove to be surprisingly difficult. Fraudsters are often unwilling to hand in their material, so how is fabrication proven without this data? Regularly, when suspicion of misconduct does arise, seemingly valid excuses are produced to explain the lack of material: 'it was a long time ago'; 'the original data was destroyed'; or 'my computer crashed'. These excuses are sometimes accompanied with authoritarian arguments, like 'who are you to criticize a tenured researcher?'. Sometimes these arguments even resort to downright threats, along the lines of 'this will destroy your career.' Facing these types of situations undoubtedly makes whistleblowing an unattractive, if not risky undertaking (Box 5.1).

5.3 Cheating

5.3.1 Cheating: A Shortcut to Knowledge?

There is good reason to consider the practice of cheating on a test to be akin to data fabrication, rather than a form of plagiarism or falsification (although, admittedly, there is an overlap between these categories – see Chaps. 3 and 6).

Just as scientific claims need to be grounded in real research findings, the results of an exam must also be based on 'real work.' Thus, cheating as a 'short cut to

Box 5.1: Self-Correction: A Dilemma

A student reaches out for help on 'r/AskAcademia', a discussion platform on the website *Reddit*. The student writes: 'I graduated three months ago and now my teacher wants to publish the paper. While most of the data is accurate and real, for some of it I made an educated guess using some economic forecast data. I was hoping I could postpone having it published until I find accurate data, but because this is an economic topic that is so new I wasn't able to do that. So what do I do? Is there a realistic chance of me being found out? Do I have him submit the paper?'

Of several dozen responses, here are four answers posted on the message board (paraphrased by us). Which one do you prefer and why?

(a) Do not under any circumstance allow that paper to be published! You have somehow missed the point that it is your professor's reputation on the line here.

(b) For the university's sake, tell your professor the truth. They will be so relieved that they didn't publish fabricated data that they will forgive you, and possibly even praise and appreciate your honesty. Everybody screws up every now and then. But we need to try to fix our screw-ups when possible.

(c) You should say something like: I revisited the analysis and I found out that I made a critical error. I'm sorry, I should have checked more completely before turning in the assignment but I'm glad I caught it before we published it.

(d) I suggest you just keep quiet and let it publish. The Chinese GDP data and a lot of developed world data is made up, twisted, or seasonally adjusted. If concerned, build in an appendix explaining how some data was created as 'line of best fit' based on your assumptions.

knowledge' is nearly synonymous with presenting a conclusion based on fictitious data – whether the outcome is correct or not makes no difference.

What exactly is cheating? Lim and See (2001) offer a list that covers a wide range of betrayals to academic integrity. To name a few, cheating can come in the form of using unauthorized material, stealing exams, lying about circumstances (to get special consideration), allowing team members to do the bulk of the work, inventing data, listing unread or even nonexistent sources, copying from a neighbor during a test, or allowing a neighbor to copy from you.

A discussion of the many tricks used by students to cheat on exams can be found in Harold Noah and Max Eckstein's instructive 2001 book *Fraud and Education: The Worm in the Apple*. The strategies they identified are far ranging and many involve a fair share of creativity; scribbling notes on their skin, tapping codes on the floor, stealing test papers, printing and attaching cheat sheets on the inside of a water bottle's label, and even sending impersonators to take tests on their behalf.

In the decades since Noah and Eckstein published their book, strategies today likely employ the services of the digital age, such as messaging apps, smart watches, and Bluetooth earbuds. Vincent Versluis and Arie de Wild (2015), of the Rotterdam University of Applied Sciences, investigated 'digital cheating' during exams in higher education and concluded that institutions seriously lag behind. Neither teachers nor administrators seemed aware of the scale or magnitude of modern forms of fraud, let alone how to counter it.

5.3.2 Dealing with Deception

Before exploring the prevalence of cheating, we must first examine a few actual cases that have come before a university board of examination. What are the common forms of cheating that universities experience and how do they respond?

Cheat sheets Recently, a student at Utrecht University was caught using the oldest trick in the book, a cheat sheet. They scribbled extensive notes and figures in their dictionary, and were caught during a routine patrol of the room. The case was reported to the board of examination, who decided to annul their exam. Furthermore, they received an official slap on the wrist that went into their record, and they were excluded from the course for a year (Figs. 5.2 and 5.3).

Falsifying grade lists A law student wanted to switch from Erasmus University to Leiden University and believed it would be necessary to falsify their grades before applying. In the process, they also forged the signature of a university employee. This was regarded as a *criminal* act when the forgeries were discovered, and the student ended up in court. Before the judge, they dramatically declared: 'I saw no other way out. It felt like either a diploma or death.' They received a suspended jail sentence of 2 weeks and 60 h community service for the forgery (Bonger, 2015).

Scheming Two students at Utrecht University developed the following scheme. Both showed up at the same exam and when it was time for submission, they got up in unison, proceeded to the examiners table, and bumped into each other 'by accident' on the way before dropping their paperwork on the floor. While they scooped up their belongings, they swapped papers, thus allowing one to hand in the exam of the other. The other, never having enrolled in this class, slipped away in the confusion unnoticed.

The scheme would have worked had two fellow students not witnessed the deceit and decided to report it to the teachers. The two were thereupon interviewed, but they categorically denied all allegations. A forensic expert was then consulted, who examined their handwriting and the allegations were confirmed. Both students were expelled from the university on account of severe academic misconduct (case reported to the author by a member of the board of examination at UU).

Photographing exams In 2012, at Tilburg University, a student was reported by several anonymous peers photographing tests with their cell phone at multiple exams, and placing the images on Facebook. When confronted with the accusations,

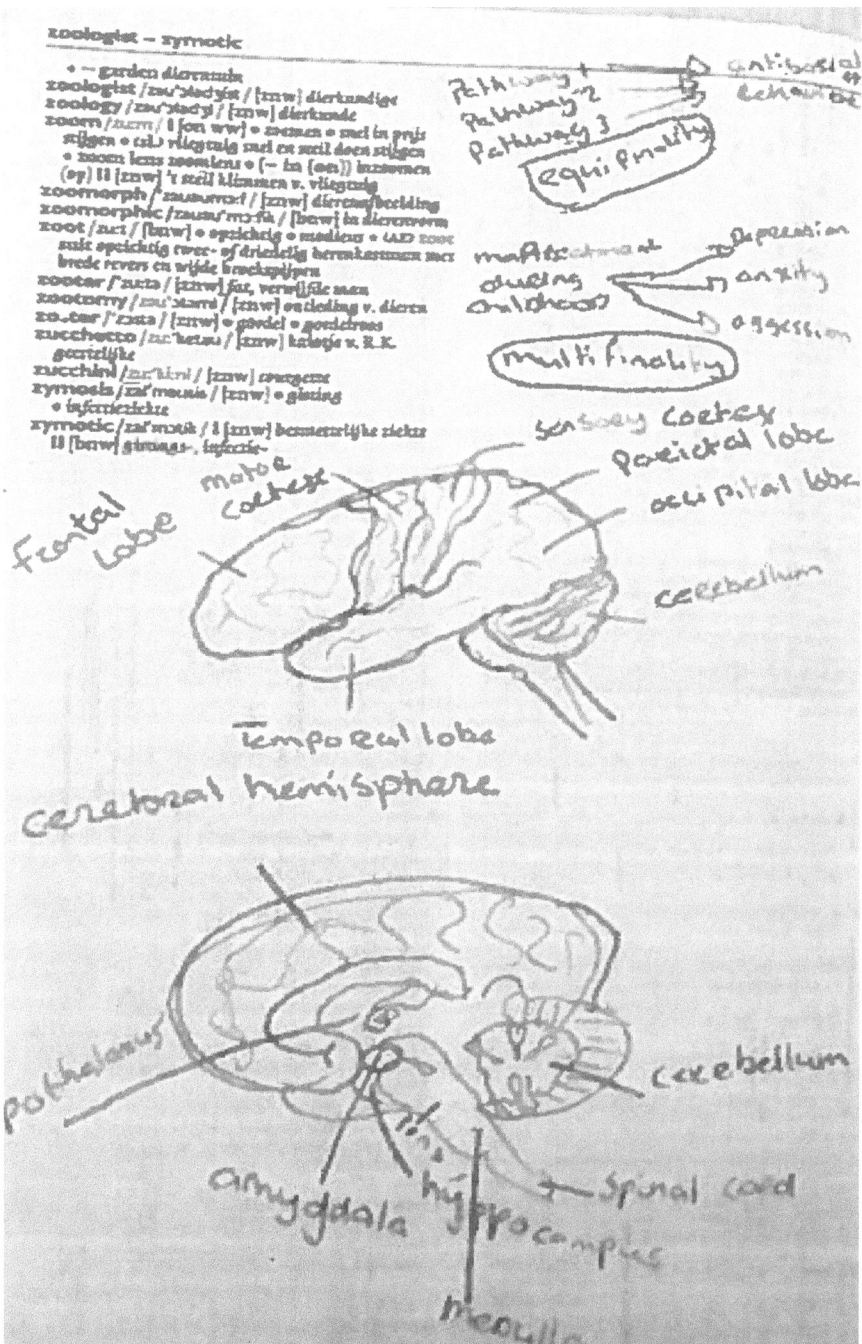

Fig. 5.2 Sheet sheet of a psychology student. Sample confiscated by a teacher and reproduced with permission of the UU board of examination

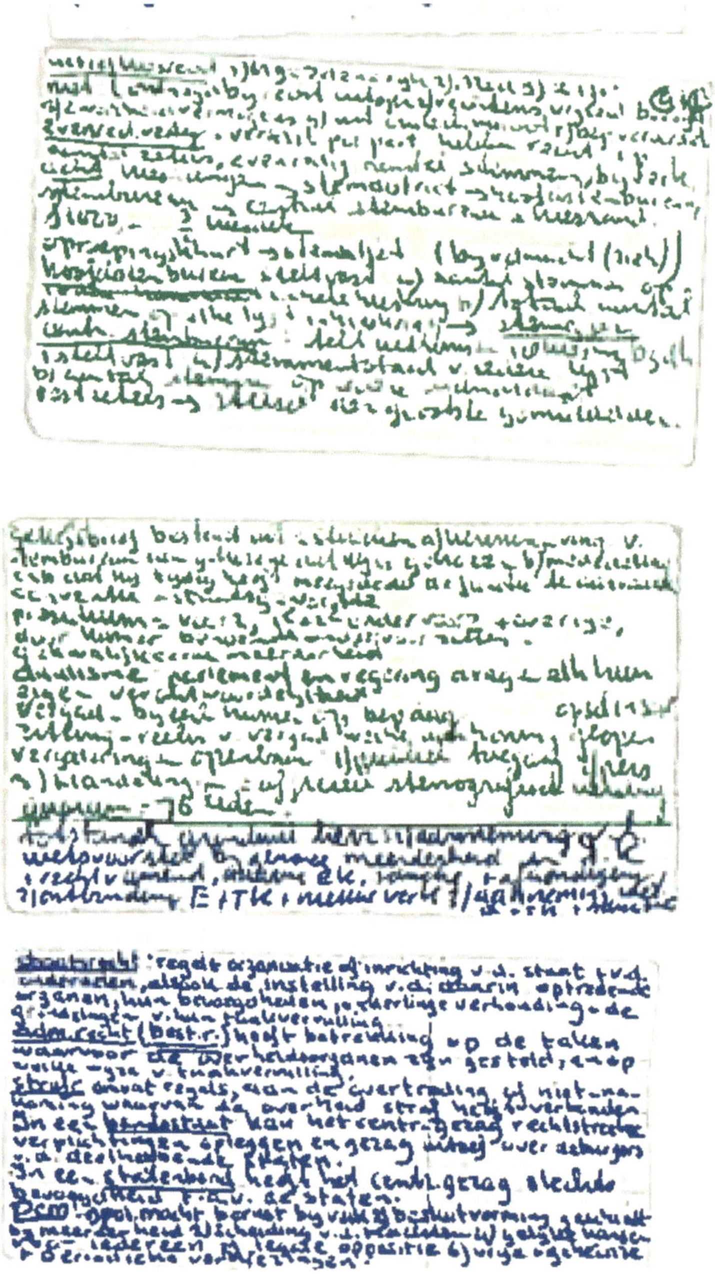

Fig. 5.3 Cheat sheets. Sample found by the author between the pages of a second-hand book. The actual size of the sheets is about 3 x 4 cm.

they confessed, maintaining that they 'had not wanted to profit from the situation, financially or otherwise'. They said they merely wanted to be able to study the questions at home, although agreeing it was wrong. The teacher, however, suspected the photos were already put up on Facebook *during* the exam, and believed the student might have been soliciting help from the outside. This could not be proven though.

The board of examination ruled that for two of her exams, where fraud could be proven, the results would be annulled. The student was furthermore excluded from all courses for the remainder of the year, as well as the entirety of the next. The student appealed against the ruling, claiming that the sentence was 'disproportional' and that they would be unable to finish their BA in time, thus facing a significant financial drawback. They additionally could also lose their position in the master's program the next year. The appeal was dismissed, on grounds of the fraud being 'extraordinarily serious' (ruling 972 of the board of examination at Tilburg University 2012).

Logging in twice In April 2014, large-scale fraud was detected during a digital exam in a statistics course for business and economy students at Amsterdam University. Students would log into the exam twice, using two different browsers. The first browser was used to work on the questions. Once the questions were answered, the digital exam revealed the correct responses. Students would then submit the now-known answers into the exam open in the second browser. Some 400 students passed the exam with abnormally high marks, which lead to suspicion of fraud. Closer inspection further revealed that the students had completed the test unrealistically fast. The exam was annulled for all 400 students (Anonymous, 2014).

Exams annulled In October 2016, some 100 pedagogy students at Salzburg University completed a test but never received the results. The entire examination was annulled after it was discovered that a number of students had discussed the multiple-choice questions used in previous exams in a closed Facebook group. Students protested against the ruling, and a discussion arose as to whether their behavior was in fact illegal. The vice-rector of the university said that copying questions and distributing them in itself wasn't wrong as long as the answers were not included. The course coordinator discovered, however, that all of his examination questions (a total of 14 pages) had been photographed by students, including the answers, and hence the annulment remained (Anonymous, 2016).

5.3.3 Is There a Cheating Crisis?

In May 2016, the *Irish Mirror,* using the Freedom of Information Act, revealed that between 2012 and 2015, over 800 students had been caught cheating across seven universities in Ireland, and that only a few students had been reprimanded, with none being expelled. 'Cheating' was broadly understood here to cover a range of exam conduct violations, including plagiarism, impersonation, and ghost writing.

Just a few months earlier, English newspapers, taking advantage of the same legislation, reported that almost 50,000 students at British universities had been caught cheating in the last 3 years. A disproportional percentage of whom, it was added, were international students from outside the EU. Similarly, in April 2016, *The Adelaide Advertizer* reported that more than 1800 students at Flinders and Adelaide Universities in south Australia had been caught copying one another's work and cheating on exams since 2010.

Has dishonest behavior among university students reached endemic proportions? Current research on academic misconduct seems to support this dramatic conclusion, at least to a certain point. Diekhoff et al. (1996) found a significant rise in (self-reported) cheating attitudes and behaviors between 1984 and 1994 in a group of students at midwestern universities in the United States. The prevalence of cheating behaviors (on exams, quizzes, or assignments) went up from 54.1% in 1984 to 61.1% in 1994. Twenty years later, Vandehey et al. (2007) repeated the study and found a slight decrease rather than increase among a similar student demographic. Overall, they concluded that instances of cheating behavior dropped to 57.4%, but strikingly, was still represented in a majority of students.

Similar trends have been described by fellow researchers in the study of academic dishonesty. McCabe and Trevino (1996) observed that self-reported admissions of academic misconduct (cheating on an exam, for example) saw a substantial increase from 39% to 64% between 1963 and 1993. While some researchers reported more conservative figures, others painted an even darker picture, claiming that only a small *minority* of students didn't engage in some form of cheating (of note, it is difficult to assess how accurate these reports are; see Franklin-Stokes and Newstead 1995). Following the findings of Anderson and Murdoch (2007), it can be safely said that cheating is both fairly common, and at the same time, seriously underestimated by teachers.

Is the 'cheating crisis' perceived universally in universities around the world? This question is hard to answer. It has been reported that, for example, post-communist central eastern countries in Europe have a higher prevalence of cheating behavior than other European countries (Pabian, 2015), and that Hong Kong business students are less likely to engage in cheating behavior than American business students (Chapman and Lupton 2004).

These isolated comparative studies of specific academic communities reveal little about the national character of academic (mis)conduct. Given the confusion over what exactly constitutes 'cheating' (see Box 5.2 for an overview of academic

---

Box 5.2: 'Classification of Forms of Academic Dishonesty'
(List Compiled from Different Encyclopedic Works)
Cheating: *Use of illegal tools, attempt to obtain external assistance during an examination, and use of unauthorized prior knowledge.*
Deception: *Providing false information to an instructor concerning a formal academic exercise—i.e., giving a false excuse for missing a deadline or falsely claiming to have submitted work.*

(continued)

Box 5.2 (continued)

Fabrication: *Presentation and reporting of fake or non-existent research data and findings.*
Facilitation: *Helping or attempting to help another commit an act of academic dishonesty.*
Ghostwriting: *Submitting work written by a third party.*
Impersonation: *Assuming another student's identity with the intent to provide the student an advantage.*
Plagiarism: *Appropriation of someone else's work (or ideas) and passing it off as one's own.*

dishonesty), the scarcity of studies into academic cheating, and the notorious unreliability of self-reporting, on which most studies are based, the exact magnitude and impact of the 'cheating crisis' will probably remain clouded for some time to come.

5.4 Ghostwriting

5.4.1 Ghost in the Machine

'Ghostwriting' is a practice that emerged in the 1980s and 1990s. David Healy from University of Wales College in Medicine, UK described one of his experiences with the phenomenon. He once received an email from a pharmaceutical company with a paper attached, its premise based on Healy's own published work. It looked as if he had written it himself; in fact, it was 'a recognizable Healy piece' (2005, p. 41). The paper was offered to him as an article he could publish under his own name. He declined on grounds that it's unethical to publish papers you haven't written yourself.

However, when a different company sent him a similar offer some 2 years later, he decided to see what would happen if he accepted but altered the content of the paper significantly. In spite of the assurance that he was 'free to edit the original article,' his changes were not accepted. Healy thereupon withdrew his name from the article. The paper, written *for* him but not *by* him, was eventually published under someone else's name.

Horace Freeman Judson reveals in *The Great Betrayal* (2004) the motives behind this type of 'ghostwriting' (presenting finished manuscripts to acknowledged scientists as a 'gift'): they are created by large pharmaceutical companies to present their products in a favorable light. The ready-made ghostwritten papers invariably report positively on a specific product (a certain drug, therapy, or medication). These papers could constitute a form of 'product placement.' They are well-written and mimic real tests and are therefore difficult to distinguish from proper research.

Ghostwriting has not only dramatically increased in frequency, it has also 'professionalized.' Sismondo (2009) writes how pharmaceutical companies now plan publications strategically in advance of the actual research taking place. Companies map out key messages, determine information relevant to various audiences and journals, and identify potential authors for their papers. Once research becomes available, 'publication planners' hire writers, negotiate with potential authors, and 'shepherd the papers through journals' submission and review procedures' (p. 175).

This may seem shady enough, but defenders of this practice claim that science is a collaborative enterprise. 'Jointly authored papers are the rule in science, not the exception, and medical writers often produce clearer, more readable papers than medical researchers themselves' (Moffatt and Elliott 2007, p. 21).

On the other spectrum of authorship, there is a second form of ghostwriting that targets unsuccessful authors. Instead of getting compensated for having their name on a paper, these authors are offered an opportunity to pay for a 'slot' in a paper written by someone else. In a publish-or-perish culture, researchers are sometimes willing to go to great lengths to keep up with the pace. *Science* reported in its issue on January 10th, 2014 on how Chinese brokers sell 'co-authorship' in papers already accepted for publication. Fees range from $1600 to $26,300, depending on the impact factor of the journal (Fig. 5.4).

To be clear, journal editors and peer reviewers do not appreciate either form of 'ghost authorship,' but are sometimes hard-pressed by industries and publishing companies to accept the practice as a fact of modern life. Thus, one publisher is quoted as saying in response to his editor's opposition of ghostwriting: 'Fine, you may have that view, but what you're actually doing is driving it underground. It's far

Fig. 5.4 *The Ghostwriter*

> **Box 5.3: 'Putting the Supervisor First: A Dilemma'**
>
> You have just finished your master's project and you want to submit your thesis as an article to a journal. Next year you plan to continue as a PhD student at the same institution. The supervisor of your master's project has announced that if you are accepted into the program, they will also be the supervisor of your PhD project. Additionally, they tell you that they want their name on your article as a first author, even though they contributed little to the project. This will improve your chances for the PhD position, they inform you. How do you respond? Choose one of the following options and prepare an argument defending your selection.
>
> 1. Ignore the request and submit the paper in your name only, running the risk of not getting accepted to the PhD program.
> 2. Accept the request and put forth the supervisor's name as a first author.
> 3. Report the incident as unethical behavior to the integrity officer.
> 4. Go to the dean of the faculty and discuss it with them first.
>
> [Adapted with permission from the Erasmus Ethical Dilemma game].

better to be transparent and get this out into the open' (quoted in Sismondo 2009, p. 181).

Ghost authorship is more dominant in the medical sciences, where the interests at stake are far greater than anywhere else. Specialized agencies offer to write research proposals for researchers at a no-cure-no-pay basis. They claim a percentage of the grant money if the application is accepted. Few would consider this 'cheating,' yet one must ask where the involvement of such ghostwriters will end. Should they be made responsible for the formulation of research questions or the development of instruments too? Wouldn't this allow them to steer research in a particular direction? (Box 5.3).

5.4.2 Hiring a Helping Hand

At various stages in their academic careers, students are sometimes confronted with 'ghost authorship' as well. Paid services offer a 'helping hand' in writing papers or preparing for exams by providing 'exercise materials.' In the last decade, commercial 'abstracting desks' have materialized in and around universities, who advertise with summaries and abstracts of course books, practice questions, and even lecture notes.

Most of these texts are written by students who get paid a nominal fee for their work, which is then offered for sale to other students. With little or no quality control, many of these texts are subpar. Despite this, there are students who claim to have successfully finished courses relying solely on these commercially produced abstracts, without even so much as opening up the course book.

Some desks offer additional 'writing guidance' to students, assuming (part of) the task of teachers, in service of helping students improve their writing skills. Others go one step further. They are called 'paper mills' or 'essay mills' and offer fully formed essays for sale. The first practice (writing guidance) is wholly legitimate, the second (paper mills) clearly isn't (Dickerson 2007).

Are students aware of the ethical dilemmas attached to these services desks? Zheng and Cheng (2015), who themselves were students when they did research for their article, interviewed peers at the University of San Francisco on their perspectives on hiring ghostwriters, and found to their surprise that a number of them (especially international students with English as a second language) did not see the practice as cheating, as long as it was only done once or twice. Some would argue that 'ghostwriting is a cooperative form of work and both parties [i.e. the student who gets paid for his or her services and the one who pays] gain mutual benefits.' Other students using ghostwriters agreed it was wrong but said in their defense that they did so because they were pressed for time, found the assessment too difficult or unclear, or just wanted a good grade.

When Zheng and Cheng subsequently interviewed a ghostwriter and asked how they felt about their work, the ghostwriter appeared not at all troubled by the ethical implications. Matter-of-factly, they remarked: 'the good thing that I've gained from this job is not just money but also the writing skill' (2015, p. 128) (Box 5.4).

Box 5.4: 'Paper Mills'

'Paper mills' or 'essay mills' are sketchy organizations that claim to offer 'original' and even 'custom-made' essays on any topic, at any level. Allegedly, they work with authors who have earned a PhD degree, or possess otherwise respectable credentials. However, many of these agencies operate in the shadows, and some are downright swindlers.

Paper mills could pose a greater threat to academic integrity than plagiarism (Thomas, 2015), and the production quality of these organizations often leaves much to be desired, at least judging from one of the clients, who goes by the nickname 'Thanatos' and has a complaint about a company called 'IVY dissertations writing services':

I used IVY thesis recently and not only did they send me a paper copied and pasted from other sources, it was the wrong paper all together! After I complained to them about the paper, they sent me a 'revised' paper on the correct subject, but again it was a simple copy and paste from 1 or 2 different websites. A simple Google search revealed they didn't even attempt to change the writing from the original websites. When I complained to IVY, they sent me a 'revised' paper that was exactly the same as the first, but now it had misspelled words and words used in the wrong context throughout the paper. I complained again, and they sent the exact same paper without any revisions made. After that point, they stopped responding to my e-mails (Thanatos, 2006).

5.4.3 Where to Draw the Line?

With regard to ghostwriting, there are two ethical issues that must be considered. One affects the individual researcher, who must decide where to draw the line. Hiring others to do your work is wrong, but on the other hand, collaboration is becoming more and more common practice in science (even though it's often not properly acknowledged, see Farrell 2001). This raises the question of who involved in the research should be granted co-authorship. Would that include fellow researchers? The project manager? Even the lab technician?

The second question affects the academic community, who has an obligation to protect objectivity and transparency. Peer review plays an important role in this obligation. It comes down to critical scrutiny for any internal and technical flaws. Some fear that ghostwriting bypasses this critical process. Virginia Barbour, chairperson of COPE (Committee on Publication Ethics), expressed her concerns that academic peer review is being subverted by 'almost industrial attempts by groups outside of normal publishing' (Barbour 2017) (Box 5.5).

Box 5.5: 'Free Riding: Misunderstood or Underreported?'
Imagine you are working with two other students on an assignment. There are certain items each of you need to work on, and the deadline is in 3 weeks. During that time, you realize that Alexandra, one of your teammates, continues to find new excuses for why she can't do her share of the work. Last week, she explains, she fell ill, before that she had to move, yesterday her computer crashed, and this morning she got into a fight with her boyfriend. When she finally does deliver, her work is inadequate. You and your teammates end up doing the lion's share of the work, even rewriting parts of Alexandra's section. When the paper authored by your team gets a 'B-', you feel cheated by Alexandra.

Most students are all too familiar with behavior like Alexandra's, known as 'free riding.' It tops the list of common student annoyances, although the 'better than average' principle seems in operation here: the free rider is always the *other* student.

Free riding is known to be demotivating even for diligent students, causing the entire team to perform suboptimally (a phenomenon called the 'sucker effect', see Swaray, 2011). Most universities recognize free riding as a negative side effect of group work and find it neither desirable nor acceptable in an academic environment. Thus the board of examination at the University of Twente declared that 'free-riding behavior, that is benefiting from other people's efforts in group assignments while not putting in the same effort as the other group members, can be considered as *fraud*' (source: www.utwente.nl/en/bms, emphasis added).

Indeed, free riding is no different than cheating and should therefore be filed under 'fraud' – but is it treated as such at universities? In general, the

(continued)

Box 5.5 (continued)

answer is no. Firstly, free riding often flies under the radar, since students (understandably) don't want to 'snitch' on their peers. Secondly, even when reported, it is difficult to prove. Alexandra, in our example, could claim that she *did* contribute to the team's effort. Is it her fault that the other two decided to rewrite her work?

Although acknowledging the problem, many universities find it hard to counter free riding. However, recently developed programs seem to be helping reduce its prevalence. Swaray (2012) reports that randomly selecting one group member to present the group's work increases participation, and cooperative learning is stimulated as a result. Further research by Maiden and Perry (2011) report that identifying individual contributions to group work is important, especially because groups can request that underperformers account for their behavior.

Romy Nefs (2019) proposed the following strategies to counter free riding:

- Use of small groups to allow for easy identification of individual contributions
- Clear assignments with well-structured schedules and strict deadlines
- Team kickoff meetings with mandatory division of labor taking place right away
- Team progress evaluation at midway point
- Evaluation of the team's work at the project's end
- Training on how to give constructive feedback to team members

5.5 Fraud Facilitating Factors

5.5.1 What Causes Fraud?

We return to the question posed at the beginning of this chapter: what circumstances or factors lead researchers to fabricate research findings (and students to cheat)? Researchers find this to be a particularly difficult question to answer, offering a host of explanations. We review four different dimensions to the problem.

5.5.2 Psychological Dimension

There are indications that scientific misconduct may be a sign of 'moral weakness,' as the virtue approach would predict it (see Chap. 3). A modern-day idiom for behaviors of 'moral weakness' could be 'anti-social personality disorder,' which is associated with irresponsible behavior, grandiose feelings of self-worth, and a general lack of guilt. An example of this can be seen in the case of fraud Cyril Burt (see Box 5.6), who was later described as a 'sick and tortured' man; the enormity of his trickery was anything but rational (Gould 1981, p. 236).

Box 5.6: 'The Case of Cyril Burt'

Educational psychologist Cyril Burt (1883–1971) has been regarded as one of the greatest frauds of the social sciences, at least until Diederik Stapel later assumed this dubious distinction.

A leading figure in his field between the 1940s and 1960s, Burt's most important research examined the heritability of intelligence. In particular, his work on monozygotic (or identical) twins was considered groundbreaking at the time. Having collected data on identical twins from 1909 to 1930, Burt used then state of the art statistics to calculate the correlation of the Intelligence Quotient (IQ) of identical twins who had been raised together and those who had not, comparing those with IQs of fraternal (non-identical) twins. Based on these findings, he claimed that intelligence has a very strong genetic driver.

In papers published between 1943 and 1966, Burt reported IQ correlations of 0.771 for identical twins raised apart, and 0.944 for identical twins raised together, fueling rhetoric that compensatory education is 'wasted money.' In the 1960s, Arthur Jensen, following Burt's lead, argued that 'for many people, there is nothing they can learn that will repay the cost of the teaching' (quoted in Tucker, 1997, p.156).

However, just months after Burt's death in 1971, Leon Kamin, a Princeton psychologist, pointed out several problems in Burt's work. For one, the number of monozygotic twins raised apart grew with every publication. Burt had started with a mere 15 pairs in 1909 and ended roughly 50 years later with 53, even though he had long since stopped collecting data. Identical twins separated at birth are a rare commodity. Additionally, Kamin found that the correlations reported remained exactly the same. He mused that the chances of finding the exact same correlation every time is close to zero. Remarkably, Kamin identified even more of Burt's foibles, finding that the two assistants he had supposedly worked with were seemingly nonexistent. Further still, it appeared Burt's data was constructed from ideal statistical distributions, rather than measured in reality (Gould, 1981, p. 235). To add insult to injury, Burt burnt his scientific papers shortly before his death, making foul play difficult to prove.

Burt's supporters attempted to explain away some of the most ostensible problems, interpreting them as 'sloppiness,' not fraud, and accusing 'left winged environmentalists' of slandering his name. However, even Burt's official biographer, Leslie Hearnshaw, who had access to his diaries, gradually came to the realization that his research was completely fraudulent. By the

(continued)

Box 5.6 (continued)

late 1970s, the verdict was accepted that Burt, once called the 'dean of the world's psychologists,' had likely fabricated most of his data. A meticulous historical analysis of the case by William Tucker (1997) showed that Burt was a fraud beyond reasonable doubt (Fig. 5.5).

Fig. 5.5 Sir Cyril Burt in the 1930s. (Source: Wikicommons)

Were frauds to be understood in terms of their psychological condition only, it would help explain why their behavior can be so reckless and self-destructive. After all, how could high profile authors producing fraudulent studies expect their deceit to go undetected? But would this explanation also help understand less serious forms of fraud (such as cheating on an exam), that people from all walks of life may commit?

Social psychologist Scott Wowra of the University of Florida probed first year psychology students at a southeastern university in the US, using an 'integrity scale' to measure the strength of their 'moral identity' (i.e. the incorporation of ideals of justice and fairness). He related student's moral identity to their ability to recall anti-social behavior, including academic dishonesty, and found a negative correlation. Thus, the 'relative centrality' of a college student's moral identity appears to affect his or her willingness to engage in academic dishonesty' (Wowra 2007, p. 317).

5.5.3   Situational Dimension

From an economic perspective, fraud in science may be all but irrational. To those seeking the highest outcome for the lowest cost, misconduct may be considered rational behavior. Economist James Wible of the University of New Hampshire argues that statistically-inclined, opportunistic scientists 'estimate the probability and the expected utility of successful evasion from discovery and then make a conscious choice to commit or not commit fraud' (1992, p. 21).

If this is true, then their decision-making depends on (a) the relative gains of committing fraud, (b) the probability of getting caught, and (c) the sort of punishments one can expect to encounter when caught. In this calculation, the chances that one will engage in fabricating data can be expected to decrease with the probability of being discovered and the weight of the penalty.

The same applies, of course, to students who may be seduced into engaging in cheating behavior when the situation appears inviting or rewarding enough. Consequently, it can be argued that a lack of reliable systems in place to monitor for cheating, unfamiliarity with university policies, and the atmosphere of secrecy that so often surrounds fraud at universities, all contribute to the continuation of the conditions that breed cheating.

5.5.4   Cultural Dimension

A third approach explaining academic fraud places emphasis on the institutional teaching and research cultures at universities. Various cultural factors have been said to influence the incidence of fraud.

One such factor is *publication pressure*. According to sociologist Patricia Woolf of Princeton University, academic 'publication is no longer just a way to communicate information. It has come to be a way of evaluating scientists; in many cases it is the primary factor in professional advancement' (1986, p. 254). In the decades since this was written, competition among universities, individual researchers, and even students has risen, as has the drive toward more scientific productivity and the call for 'excellence.' Some argue that this pressure caused scientists to cut corners (see; Fanelli, 2010a, b, 2012). We return to this issue in Chap. 9.

Another factor is *peer culture*, the pressure one feels to conform to the prevailing attitudes of their peers. In a survey of US college students, Rettinger and Kramer found that decisions to cheat depended at least partly on one's perception that others were cheating too. They concluded that 'seeing cheating is the beginning of a social learning process. New students learn how to behave by observing their peer(s)' (2009, p. 310). More particularly, performance-oriented teaching styles in class, coupled with poor instruction, can lead students to justify cheating (Murdock in Rettinger and Kramer 2009). Similarly, Shu Ching Yang (2012, p. 235), who examined academic dishonesty among Taiwanese students, found that the behaviors and attitudes of peer groups influenced student decision making regarding such conduct.

5.5.5 *Integrative Perspective*

Various attempts were undertaken to integrate personal, situational, and cultural dimensions into a unified model for analyzing cases of fraud. One such model, presented by Donald Cressey, is coined as the 'Fraud Triangle' (1973). It combines three factors: incentives to commit fraud ('opportunity'), various contextual factors ('pressure'), and the perception of an action as fitting into one's personal code of ethics ('rationalization'). Thus, when students claim to be unclear about what behaviors constitute academic dishonesty or say a particular course 'isn't relevant for their future career,' they *rationalize*. When they cite increased competition for academic positions, they perceive *pressure*. And when they make use of a gap in an exam's procedures, they take advantage of an *opportunity* (Hayes et al. 2006).

Becker, Connoly, Lentz, and Morison (2006) found that all three factors predict dishonest behavior in business students (who rank the most likely to cheat). Their conclusion was largely confirmed by Choo and Tan (2008), who also identified that the three factors all held influence on a student's propensity to cheat (Figs. 5.6 and 5.7).

Breaking the Fraud Triangle (opportunity, pressure, rationalization) is regarded as a key to its deterrence. Since the three elements strongly interact, removing one would significantly reduce the risk of unethical behaviors emerging. Of the three, opportunity is 'most directly affected by the system of internal controls and generally provides the most actionable route to deterrence of fraud' (Cendrowski, Martin, & Petro, *The Handbook of Fraud Deterrence*, 2007, p.41) (Box 5.7).

Fig. 5.6 *Fraud Triangle.*
(After Cressey, 1973)

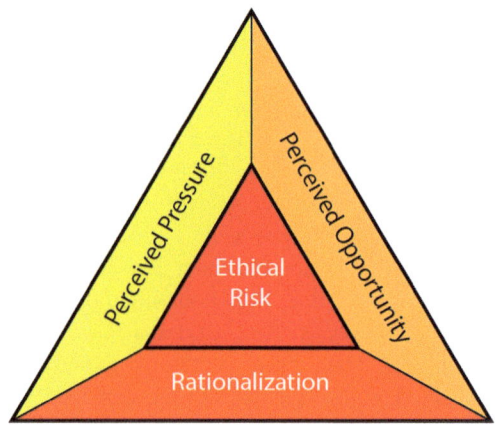

Fig. 5.7 'Alex, why are you so stressed out?'. (Photo cartoon © Ype Driessen, 2019, reproduced with permission from the author)

Box 5.7: 'Hoaxing'

A 'hoax' is a prank, a small con committed on an individual or group of people, who are made to believe something only to find out that the joke's on them. Hoaxes typically involve the production of some form of falsehood, but they aren't classified as 'fraud' because the intention is not to profit from the deceit.

The notorious 1996 'Sokal Hoax' was a practical joke played on French postmodernist sociologists and their followers. Alan Sokal, professor of physics at New York University, composed a text, entitled 'Transgressing the Boundaries: Towards a Transformative Hermeneutics of Quantum Gravity,' which was made up largely of (attributed as well as non-attributed) quotations from prominent French postmodernists, including, to name a few, Gilles Deleuze, Jacques Derrida, Jacques Lacan, and Bruno Latour. In the paper, Sokal argued that 'physical "reality", no less than social "reality" is at the bottom a social and linguistic construct' (Sokal and Bricmond 1998, p. 2).

(continued)

Box 5.7 (continued)

Sokal submitted the article to *Social Text*, a leading American cultural-studies journal, despite believing it to be complete gibberish and full of logical errors. Shortly after *Social Text* accepted the article and ran it in the Spring 1996 issue, Sokal came out, declaring it a 'parody.' He proclaimed that his intentions were to expose postmodernist discourse as pretentious drivel. Sokal argued that despite frequent references to subjects like quantum mechanics, string theory, and Einstein's general theory of relativity, postmodernists possessed a completely flawed understanding of the natural sciences. A follow-up book, entitled *Fashionable Nonsense. Postmodern Intellectuals' Abuse of Science* (Sokal and Bricmond 1998) spelled out his argument in further detail.

Following in the footsteps of Sokal, Peter Boghossian and two of his colleagues at Portland State University carried out a similar but more elaborate hoax in 2018, known as the 'grievance studies affair.' They wrote no less than 20 articles, promoting deliberately absurd ideas on morality and morally questionably acts, and submitted them to journals on post-colonial theory, gender studies, queer theory, and intersectional feminism – which they dubbed 'grievance studies' because in these fields 'grievances are put ahead of objective truth.' Seven articles were accepted (four even published), nine were rejected, and the remaining were under review or in the process of resubmission when the hoax was revealed.

The hoax, aimed to expose the lack of scientific rigor in postmodern research, backfired when Boghossian and his colleagues were critiqued for the same reason, as they had not included a control group in their experiment, and even had to face a research misconduct inquiry on the grounds of conducting human subject-based research without approval, and for fabricating data.

Hoaxes such as these are not just 'practical jokes.' They are meant to be critiques of scientific practices, directed at the shortcomings of quality control in the publishing process, and purported to raise awareness of the lack of critical faculties in some academic circles. However, they also raise questions themselves. Is it, for example, ethically acceptable to waste recourses in this way? And do these authors not act in bad faith, deliberately misrepresenting the fields of research they purport to expose?

5.6 Clearing Science: Measures to Counter Fabrication

5.6.1 Fake Science

Fabrication is a form of academic misconduct that belongs in the realm of 'fake science.' It is a deliberate attempt at deceit. What can be done to counter it? We discuss three general strategies.

5.6.2   Academic Peers

An important role in exposing fraud is reserved to the academic community. Both Stapel and Burt were unmasked by fellow researchers, cheating students by their peers. But whistleblowing is an unappealing option as we have seen, and not always appreciated in the academic community.

Lim and See (2001) report that Singaporean students are quite tolerant of academic dishonesty, with the majority of them preferring to ignore the problem rather than report it. One student commented: 'Nobody will report another student for cheating as you may be the one cheating someday' (p. 272). Malgwi and Rakovski (2009) found that American students were just as reluctant to report academic dishonesty to the relevant authorities, and preferred other counter measures (including stronger penalties, parental notifications, or use of an anonymous tip line).

5.6.3   Proctoring or Disciplining?

Many notorious frauds were in the position to hide their actions. Would putting more checks in place, and not allowing the opportunity to fabricate data in the first place play a role in diminishing the ethical risks?

With forms of online and distance learning rapidly expanding at universities, 'proctoring' (supervising students taking exams, verifying their identities, and other forms of vigilance) becomes indispensable to not 'giving an opportunity' to cheaters.

Research by Prince, Fulton, and Garsombke (2009) suggests that some form of vigilance is justifiable, but it can easily transform into ludicrous distrust, as the 'Classroom Management and Student Conduct' page of *WikiHow* reveals. On the page with tips for teachers, we find such suggestions as this: *Greet the student as they come into the classroom, look them in the eye, and watch for signs of nervousness, while simultaneously inspecting their arms to see if notes are written on them.* Also: *Know that some female students might write on their legs but be aware that that observing this behavior might lead to an accusation of harassment.*

In a climate of mistrust and suspicion, students will complain that campus integrity policies are biased against them (see McCabe 2005). Or worse, argues Zwagerman (2008, p. 6909), in a climate that is entirely designed to eliminate every opportunity to cheat, suppression of academic dishonesty becomes 'more important than anything that might be sacrificed in the effect – including education.'

5.6.4   Sanction or Honor Code?

Would it help to decrease incidents of cheating by increasing the penalty? In an examination of several classical cases of fraud, Bridgestock (1982) argues to the contrary. He observed that for many offenders, career pressure, or even an unusual commitment to a certain set of ideas, overrides considerations of ethics. Sanctions are 'at best a partial deterrent to fraud' (pp. 378–9).

Stephen Davis (1993) corroborates this finding. Confronted with the question: 'If a professor has strict penalties and informs the class about them at the beginning of the semester, would this prevent you from cheating?' some 40% of male students responded 'no.' Female students were only slightly more responsive. Closer examination of the data showed that the majority of the students who responded with a 'no' had reported previously cheating in college. In short: 'if students have cheated in the past and plan to cheat again, there is precious little that will sway their course of action' (1993, p. 28).

On the other hand, would an approach that capitalizes on fairmindedness and justice help? Can cheating be deterred if students are made more familiar with academic integrity, and offered an honor code to abide by? Jordan (2001) finds that indeed, non-cheaters have a greater understanding of institutional policies than cheaters do, but since cheaters and non-cheaters received the same information, the difference between them seems to lie in their attitude towards it.

McCabe advocates for a 'just community approach,' which cherishes democratic values and promotes moral reasoning. He further adds that it's not just students that need to be enlightened: 'the real key to building and sustaining an atmosphere of student integrity on any campus may be involving all members of the campus community – students, faculty, and administration' (1993, p. 656).

5.7   Conclusions

5.7.1   Summary

This chapter dealt with a wide variety of a very serious form of fraud in science, namely the fabrication of data, research findings, and test results, which can be accomplished in a number of ways. Well-known cases of forgery from the likes of disgraced academics Cyril Burt and Diederik Stapel were discussed, and the question of how to identify and expose these frauds was explored. On the flip side, the fate of those who do the exposing, the 'whistleblowers,' saw our attention.

Cheating among students, as a fraudulent 'shortcut to knowledge' is discussed, and examples of cheating are presented. From this, we examined whether cheating has increased over the years and if there was in fact a 'cheating crisis,' as some proclaim.

Furthermore, debates were presented on the practice of having others write your papers, hoaxes as a specific form of fabrication and whether they have a cleansing function, and how institutional mechanisms can help liberate academia from fraud.

5.7.2 Discussion

Cheating, fabrication, and forging of research data, among other forms of fraud, have plagued science from its humble beginnings, but has it increased in the past few decades? Is there truly a 'cheating crisis,' perhaps even beyond academia?

There are certainly indications that such a crisis exists, but at the same time, the scientific community appears more concerned with research ethics than ever before. From this, we identify two important questions to ponder. Are we doing enough to prevent or at least combat this crisis? And have conditions in science changed such that fabrication has become more lucrative or attractive? Both questions will be the subject of further discussion in subsequent chapters.

Case Study: The Temptations of Experimental Social Psychology

Ruud Abma

At the end of August 2011, social psychologist and dean of the social science faculty at Tilburg University, Diederik Stapel, was confronted with allegations of fraud. The evidence was gathered by three of Stapel's junior colleagues, who had tried in vain to replicate the results of his earlier studies. A week later, Stapel confessed to the fabrication of his data. He was immediately fired, and a committee was installed to investigate all of his publications.

As may have been expected, the media extensively covered the scandal. Psychologists immediately responded by emphasizing that Stapel was an exception and that psychological research resoundingly conformed to the rules of scientific integrity. Other scientific community members, including methodologists, pointed out the presence of a whole 'grey area' between accepted types of data cleaning (i.e. removing outliers) and outright scientific misconduct. In this grey area, they argued, a vast number of researchers fell prey to the temptation of bending the rules, using procedures such as 'cherry picking' (only reporting significant outcomes) and 'data cooking' (presenting and using processed data as raw data).

Of course, scientific misconduct is not limited to social psychology or the social sciences. It is a challenge for science itself. Around the same time Stapel's fraud was revealed, a professor in vascular surgery was found to have faked results on a massive scale. This too caused a stir in the media, but nothing compared to Stapel's case. Apparently, there is something about (social) psychology that generates an extraordinary amount of media attention (Fig. 5.8).

To understand what Stapel did, it's important to know how he worked. Together with either a colleague or a PhD student, Stapel would propose a research theme and hypothesis, and construct an experimental design. Subsequently, he would volunteer to test it – by himself. He then would return with the data already ordered into

Fig. 5.8 Diederick
Stapel's autobiographical
account of his downfall,
Derailed ('Ontspoord'),
published in 2012

PROMETHEUS

Fig. 5.8 Diederick Stapel's autobiographical account of his downfall, *Derailed* ('Ontspoord'), published in 2012

neat tables, with readymade statistical analyses in hand. From there, he instructed his PhD students to integrate them into their research articles.

On October 31st 2011, the Levelt Committee published a preliminary report (see https://www.commissielevelt.nl/) concluding that at least 30 journal publications (co)authored by Stapel had been based on fabricated data, and that these fraudulent practices had been going on since at least 2004. Thereupon, Stapel withdrew his doctoral degree and said in a public statement: 'I have used improper means to produce attractive results. In modern science, the level of ambition is high and the competition for scarce means is huge. During the last years, this pressure has gotten the better of me. I did not cope adequately with the pressure to score, to publish (…). I wanted too much too fast. In a system where there is a lack of control, where people usually work alone, I have taken a wrong turn.'

In his statement, its striking that Stapel refers to flaws in the system (pressure to publish, lack of control, etc.). This is in line with his inaugural lecture at Tilburg University in 2008, where he proclaimed: 'It is the context that determines whether you cheat or not. You cheat when you're angry, if the game lets you, if you don't want to lose from your older brother, or if you play against your six year old daughter and do not want to win.'

So, what then about the 'game' of social psychology? What is the context here that determines whether you cheat or not? According to Stapel, it was publication culture: 'Like the consumer that sees bargains and shopping-streets everywhere, the scientist that is rewarded to the publication sees potential articles everywhere. In the long run, this strategic behavior is not in the interest of the forum of science. It leads to scientific pornography, the result of a quick climax. It leads to trendy and conformist science at the expense of originality and creativity.'

It seems there were two sides to Diederik Stapel: one that conformed to the high impact publication culture, with trendy articles that guaranteed attention in both scientific journals and the popular media, and another that rebelled against this culture in the name of quality of content and long term satisfaction – corresponding to the intrinsic value of science. The inner friction between the two may have stirred the anger and cynicism that would eventually lead professor Diederik Stapel toward scientific misconduct.

When Stapel's scheme was finally found out, the self-purifying capacities of the scientific community were hailed. But the damage to the reputation of social psychology was enormous and led to both debates and policy measures within the field (Van Lange, Buunk, Ellemers, & Wigboldus, 2012) and fundamental thoughts about the status of theory formation in social psychology (Ellemers, 2013). The damage he brought on his colleagues and PhD students, who had contributed to his research and publications but were kept in the dark about his fraudulent practices, was immeasurable: apart from the emotional shock resulting from such a severe breach of trust, they had to re-evaluate their publication list and ward off suspicions about their own conduct.

How was it possible that Stapel was able to skirt detection for possibly more than 15 years? Does this mean that the structures of (social) science, with peer review at its foundation, are starting to crumble? The Levelt-reports presented an astonishing example of peer review failing to act as a check on bad science. And that is exactly what the committee's statistical experts found throughout Stapel's publications; 'very doubtful results', 'highly implausible results', 'unbelievably high factor loadings', and 'results extremely unlikely', to name a few.

The scientific fraud perpetrated by Stapel prompted authorities in academia to introduce stricter rules and regulations, and increased efforts to better inform students of the misgivings of scientific misconduct. The Stapel case has also sparked debates on publishing habits and research subject selection, including replication of experimental studies. Compared to the allure of producing novel results, replication is not 'sexy,' and most journals are not interested in publishing replication research (see Chap. 9 for further discussion).

What is to be learned from all of this?

First of all, even if Stapel's misconduct is an isolated case – which it probably is not – his long-lasting misconduct can be seen as partly the result of an unfortunate combination of perverse incentives (publish or perish) and lack of scrutiny (by colleagues and peer reviewers).

Second, this unfortunate combination endemic within – at least – the field of social psychology is influencing a great number of researchers. Just observing the absurdly long lists of publications showcases how much researchers associate scientific quality with number of publications. As long as this system of reward and promotion via publication frequency maintains its grip on researchers, the risk of sliding down the slope from 'data cleaning' to 'data falsification' remains a distinct possibility. We end with the words of philosopher of science David Hull (1998, p. 30): 'Melodramatic as allegations of fraud can be, most scientists would agree that the major problem in science is sloppiness. In the rush to publish, too many corners are cut too often.'

Assignment

1. There is apparently something distinct about (social) psychology that allows it to generate an extraordinary amount of media attention. What could that be?
2. How do you believe Stapel was able to commit fraud for so long?
3. What could be done to prevent cases like this from happening again?

Case Study: Unexplained or Untrue?

Parapsychology is the study of psychological processes that are 'not yet understood' by the 'regular sciences.' Think of extra-sensual perception, or telepathy (thought transference). Research into these processes raise several questions, such as: How can one have access to another's thoughts without the need to communicate? Is it even possible?

William James and Gerard Heymans, the 'founding fathers' of academic psychology in the United States and the Netherlands, respectively, were among the first to take parapsychological phenomena seriously. It was Heymans who attempted to conduct proper experiments in telepathy (Heymans, Brugmans, & Weinberg, 1920). He asked a non-professional 'medium' (an individual claiming psychic sensitivity) to take part in a series of tests lasting several months, running a total of 157 trials. These experiments are regarded as some of the most successful parapsychological studies ever undertaken (Fig. 5.9).

The medium that took part in Heymans experiment was a 23-year-old mathematics student named Abraham van Dam. Van Dam was selected because he claimed the capacity to 'know' the whereabout of hidden objects by simply 'sensing' their location. He was first blindfolded, then seated in a blackened cardboard box approximately the size of a telephone booth. Inside, van Dam was able to stick his hand through a small hole in the front panel of the box. Opposite the box, and out of van Dam's line of sight, lay a rectangular board with 48 marked fields, not unlike a chess board (the fields were numbered A1, A2, etc.).

When van Dam was seated in the booth, Heymans withdrew to the attic above the test room, where he could observe van Dam's hand hovering over the board. Heymans would now concentrate on a pre-selected field on the board. He would then attempt to 'steer' van Dam's hand to the correct field by means of telepathy. Van Dam, being the 'receiving party.' would allow his hand to be 'conducted' by the thoughts of the other. Whenever van Dam believed he had reached the field Heymans supposedly directing him toward, he would tap with his finger on the board, thus selecting a particular field. Upon van Dam's queue, Heymans would select a new field and the same procedure would start anew. Heymans and his assistants were able to record the accuracy of van Dam's selections over a large number of attempts (Fig. 5.10).

The outcome of this experiment was truly impressive. Van Dam produced no less than 55 hits (a success ratio of 35%). This is a significant result, as the probability

Fig. 5.9 Gerard Heymans,
1857–1930. Founder of
modern Dutch psychology,
researchers of thought
transference. (Photo:
A.S. Weinberg)

Fig. 5.10 Test set up of
the 1919 telepathy
experiments. (Source:
Heymans et al. 1920, p. 4)

predicted no more than 4 hits in a series of 157 trials, a 1 in 48 chance to pick the
correct field in any given attempt. Interestingly, van Dam performed much better at
the beginning of the experiment than at the end, leading Heymans to believe that the
'medium' had gradually lost his 'psychic abilities.' Therefore, Heymans decided to

discontinue the series. Finally, it is worth noting that there were two experimental conditions: in one, the sender (Heymans or one of his co-workers) was nearby (staying in the same room), in the other, the sender was further away (staying in the attic). Remarkably, van Dam did slightly *worse* in the 'nearby condition,' suggesting there were no immediate cues he could use to determine his choice.

Historians of science have struggled to interpret this study. Proponents of parapsychology often cite the experiments as solid proof that telepathy exists. Heymans himself claimed that 'the existence of thought-transference [...], has been put beyond reasonable doubt by these experiments.' Very rarely do studies in the field of parapsychology result in such convincing evidence. But what does that mean – to have 'convincing evidence'?

In 1979, a team of researchers at Groningen University attempted to replicate Heymans' experiment using the same design, but with a different medium and modern equipment, like video cameras (Draaisma, 1970). They were resolute in ensuring that no form of 'information leakage' help the medium guess correctly.

What they found was in effect the opposite of Heymans' findings. During the entire series of trials, their medium never produced a single hit. Now, a failed replication doesn't necessarily prove that the original experiment was 'corrupted,' the researchers admitted, but it does lay barren the challenges of studying parapsychology scientifically. That is to say: if the phenomenon you study does not possess an observable set of rules, and rather its study depends on finding someone who happens to have 'psychic abilities,' then this 'inexplicable process' cannot be studied scientifically. Therefore, parapsychology cannot be considered a science.

Psychologist Hans Linschoten (1959) observed, however, that parapsychology *can* be a science if it 'distances itself entirely from any inkling towards the mystic' and its practitioners dedicate themselves to 'finding law-like psychological functions.' Inevitably, this recommendation will burden parapsychology with a terrible dilemma. To become reputable (and thus be fully accepted as a science), parapsychology must prove that its object of study (the paranormal) does not really exist, and is in fact a subset of psychology proper. But then of course, it would no longer be *para*psychology, but 'just' psychology.

Assignment

Consider the value of evidence (Heymans' findings) and counterevidence (Groningen research findings) in controversial research. When should you be prepared to accept 'strange findings,' and at what point do you accept that counterevidence has subverted your initial 'strange' findings? Replication is the key concept here but consider how difficult it could be to replicate an original study in parapsychology! Think about the relevance of the Heymans studies when contemplating the following questions:

- What does it take to decide whether or not something is an 'unexplained phenomenon' or is just 'untrue'; i.e. a chance finding?

- Should researchers be allowed to ask 'strange questions'? In other words: can the social sciences study 'elusive phenomena' such as telepathy, and if yes, are there any limits?
- Are there subjects in the social sciences that you believe resemble parapsychological phenomena in their inability to be studied scientifically?

Suggested Reading

Horace Freeman Judson's (2004) *The Great Betrayal* and John Grant's (2008) *Corrupted Science* present highly approachable accounts of both well-known and lesser-known cases of scientific fraud. *Fraud and Education: The Worm in the Apple*, by Harold Noah and Max Eckstein (2001) focuses on cases of cheating, test tampering, and other forms of professional misconduct in science. Daniele Fanelli's 2009 studies are a must read for any academic interested in these issues.

References

Anderson, E. M., & Murdoch, T. B. (2007). *The psychology of academic cheating*. London: Elsevier.

Anonymous. (2014, April 15). Grootschalige fraude dor eerste jaars economie UvA [Large-scale fraud by economy freshmen at Amsterdam University]. *Algemeen Dagblad*. Retrieved from: https://www.ad.nl/binnenland/grootschalige-fraude-door-eerstejaars-economie-uva~a62acb79/

Anonymous. (2016, December 16). Wirbel an der Uni Salzburg: 100 Prüfungen nicht beurteilt [Fuss at the University of Salzburg: 100 exams not decided]. *Salzburger Nachrichten*. Retrieved from: https://www.sn.at/salzburg/politik/wirbel-an-der-uni-salzburg-100-pruefungen-nicht-beurteilt-604705

Barbour, V. (2017, May). 'From the outgoing chair'. COPE newsletter, 5(5).

Becker, D., Connoly, J., Lentz, P., & Morison, J. (2006). Using the business fraud triangle to predict academic dishonesty among business students. *Academy of Educational Leadership Journal, 10*(1), 37–54.

Board of examinations of Tilburg University, ruling 972. (2012, July 2). Retrieved from: https://www.tilburguniversity.edu/sites/default/files/download/ID%20972_2.pdf

Bongers, V. (2015, March 19). Taakstraf voor fraude met cijferlijst. [Community service for fraud with grade list]. *Leids universitair weekblad MARA, 38*(23). Retrieved from: http://archief.mareonline.nl/archive/2015/03/18/taakstraf-voor-fraude-met-cijferlijst

Bridgestock, M. (1982). A sociological approach to fraud in science. *The Australian and New Zealand Journal of Sociology, 18*(3), 364–383. https://doi.org/10.1177/144078338201800305.

Cendrowski, H., Martin, J. P., & Petro, L. W. (2007). *The handbook of fraud deterrence*. Hoboken: J.P. Wiley & Sons.

Chapman, K. J., & Lupton, R. A. (2004). Academic dishonesty in a global educational market: A comparison of Hong Kong and American university business students. *International Journal of Educational Management, 18*(7), 425–435. https://doi.org/10.1108/09513540410563130.

Choo, F., & Tan, K. (2008). The effect of fraud triangle factors on students' cheating behaviors. In B. N. Schwartz & A. H. Catanach (Eds.), *Advances in accounting education* (Vol. 9, pp. 205–220). Bingley, UK: Emerald. https://doi.org/10.1016/S1085-4622(08)09009-3.

Commission Levelt. (2012). Failing science. In *The fraudulent research practices of social psychologist Diederik Stapel* [Falen wetenschap: De frauduleuze onderzoekspraktijken van sociaal-psycholoog Diederik Stapel]. Tilburg: Tilburg University.

Cressey, D. R. (1973). *Other people's money*. Montclair: Patterson Smith.

Crocker, J. (2011). The road to fraud starts with a single step. *Nature, 479*, 151. https://doi.org/10.1038/479151a.

Davis, S. (1993). *Cheating in college is for a career: Academic dishonesty in the 1990s.* Paper presented at the Annual Meeting of the Southeastern Psychological Association (39th, Atlanta, GA, March 24–27, 1993). Retrieved at: https://files.eric.ed.gov/fulltext/ED358382.pdf

Dickerson, D. (2007). Facilitated plagiarism: The saga of term-paper mills and the failure of legislation and litigation to control them. *Villanova Law Review, 52*(1), 21–46. https://digitalcommons.law.villanova.edu/vlr/vol52/iss1/2.

Diekhoff, G. M., Labeff, E. L., Clark, E., Williams, L. E., Francis, B., & Haines, V. J. (1996). College cheating: Ten years later. *Research in Higher Education, 37*(4), 487–502. https://doi.org/10.1007/BF01730111.

De Vrieze, J. (2017, September 23). 'Toen deze promovenda een fraudeur ontmaskerde, keken haar collega's weg' [When this PhD student unmasked a fraud, her colleagues looked the other way] *De Groene Amsterdammer*.

Fanelli, D. (2009, May 29). How many scientists fabricate and falsify research? A systematic review and meta-analysis of survey data. *PLOS One.* https://doi.org/10.1371/journal.pone.0005738

Fanelli, D. (2010a, April 7). Positive results increase down the hierarchy of the sciences. *PLoS One.* https://doi.org/10.1371/journal.pone.0010068

Fanelli, D. (2010b, April 21). Do pressures to publish increase scientists' bias? An empirical support from US states data. *PLOS One.* https://doi.org/10.1371/journal.pone.0010271

Fanelli, D. (2012). Negative results are disappearing from most disciplines and countries. *Scientometrics, 90*, 891–904. https://doi.org/10.1007/s11192-011-0494-7.

Farrell, M. P. (2001). *Collaborative circles. Friendship dynamics & creative work*. Chicago: The University of Chicago Press.

Festinger, L. (1954). A theory of social comparison process. *Human Relations, 7*(2), 117–140. https://doi.org/10.1177/001872675400700202.

Franklin-Stokes, A., & Newstead, S. E. (1995). Undergraduate cheating: Who does what and why? *Studies in Higher Education, 20*(2), 159–172. https://doi.org/10.1080/03075079512331381673.

Gould, S. J. (1981). *The mismeasurement of man*. London: Penguin.

Grant, J. (2008). *Corrupted science. Fraud, ideology and politics in science*. Wisley: Fact, Figures & Fun.

Hayes, D., Hurtt, K., & Bee, S. (2006). The war on fraud: Reducing cheating in classroom. *Journal of College Teaching & Learning, 3*(2), 1–12. https://doi.org/10.19030/tlc.v3i2.1742.

Healy, D. (2005). Shaping discontent. The role of sciences and marketing. In P. Pietikainen (Ed.), *Modernity and its discontent. Sceptical essays on the psychomedical management of malaise* (pp. 33–48). Stockholm: Axel and Margaret Ax:son Johnson Foundation.

Jordan, A. E. (2001). College cheating: The role of motivation, perceived norms, attitudes, and knowledge of institutional policy. *Ethics & Behavior, 11*(3), 233–247. https://doi.org/10.1207/S15327019EB1103_3.

Judson, H. F. (2004). *The great betrayal. Fraud in science*. Orlando: Harcourt.

Lim, V. K. G., & See, S. K. B. (2001). Attitudes toward, and intentions to report, academic cheating among students in Singapore. *Ethics & Behavior, 11*(3), 261–274. https://doi.org/10.1207/S15327019EB1103_5.

Maiden, B., & Perry, B. (2011). Dealing with free-riders in assessed group work: Results from a study at a UK university. *Assessment & Evaluation in Higher Education, 36*(4), 451–464. https://doi.org/10.1080/02602930903429302.

Malgwi, C. A., & Rakovski, C. C. (2009). Behavioral implications and evaluation of academic fraud risk factors. *Journal of Forensic & Investigative Accounting, 1*(2), 2–37.

Markowitz, D., & Hancock, J. (2014). Linguistic traces of a scientific fraud: The case of Diederik Stapel. *PLoS One, 9*(8), e105937. https://doi.org/10.1371/journal.pone.0105937.

McCabe, D. L. (1993). Faculty responses to academic dishonesty: The influence of student honor codes. *Research in Higher Education, 34*(5), 649–650. https://doi.org/10.1007/BF00991924.

McCabe, D. L. (2005). It takes a village. Academic dishonesty & educational opportunity. *Liberal Education, 91*(3), 26–31.

McCabe, D. L., & Trevino, L. K. (1996). What we know about cheating in college: Longitudinal trends and recent developments. *Change: The Magazine of Higher Learning, 28*(1), 28–33. https://doi.org/10.1007/BF00991924.

Moffatt, B., & Elliott, C. (2007). Ghost marketing: Pharmaceutical companies and ghostwritten journal articles. *Perspectives in Biology and Medicine, 50*(1), 18–31. https://doi.org/10.1353/pbm.2007.0009.

Nefs, R. (2019). *Free-riding among students: Causes, consequences and solutions*. Unpublished report, Utrecht University.

Noah, H. J., & Eckstein, M. A. (2001). *Fraud and education. The worm in the apple*. Lanham: Rowman & Littlefield.

Pabian, P. (2015). Why 'cheating' research is wrong: New departures for the study of student copying in higher education. *High Education, 69*(5), 809–821. https://doi.org/10.1007/s10734-014-9806-1.

Prince, D. J., Fulton, R. A., & Garsombke, T. W. (2009). Comparisons of proctored versus non-proctored testing strategies in graduate distance education curriculum. *Journal of College Teaching & Education, 6*(7), 51–62. https://doi.org/10.19030/tlc.v6i7.1125.

Rettinger, D. A., & Kramer, Y. (2009). Situational and personal causes of student cheating. *Research in Higher Education, 50*(3), 293–313. https://doi.org/10.1007/s11162-008-9116-5.

Singal, J. (2015, May 29). The case of the amazing gay-marriage data: How a graduate student reluctantly uncovered a huge scientific fraud. *New York Magazine*.

Sismondo, S. (2009). Ghost in the machine: Publication planning in the medical sciences. *Social Studies of Sciences, 39*(2), 171–198. https://doi.org/10.1177/0306312708101047.

Sokal, A., & Bricmond, J. (1998). *Fashionable nonsense. Postmodern intellectuals' abuse of science*. New York: Picador.

Swaray, R. (2012). An evaluation of a group project designed to reduce free-riding and promote active learning. *Assessment and Evaluation in Higher Education, 37*(3), 285–292. https://doi.org/10.1080/02602938.2010.531246.

Thanatos. (2006, November 10). *Re: Ivy dissertations is a total scam*. Retrieved from https://essay-scam.org/forum/es/ivy-dissertation-total-two-others-try-72/

Thomas, A. (August 19, 2015). 'Forget plagiarism: there's a new and bigger threat to academic integrity'. The Conversation.com/uk.

Tucker, W. H. (1997). Re-reconsidering Burt: Beyond a reasonable doubt. *Journal of the History of the Behavioral Sciences, 33*(2), 145–162. https://doi.org/10.1002/(SICI)1520-6696(199721)33:2<145::AID-JHBS6>3.0.CO;2-S.

Vandehey, M., Diekhoff, G. M., & LaBeff, E. E. (2007). College cheating: A twenty-year follow-up and the addition of an honor code. *Journal of College Student Development, 48*(4), 468–480. https://doi.org/10.1353/csd.2007.0043.

Versluis, V., & De Wild, A. F. (2015). Actieve borging tegen fraude [Active security against fraud]. *Kenniscentrum Business innovatie, Hogeschool Rotterdam*. Retrieved at: https://www.hogeschoolrotterdam.nl/onderzoek/projecten-en-publicaties/pub/actieve-borging-tegen-fraude/f6b3626e-a5e4-42b8-9199-3f2dcd469647/

Wible, J. R. (1992). Fraud in science: An economic approach. *Philosophy of the Social Sciences, 22*(1), 5–27. https://doi.org/10.1177/004839319202200101.

Woolf, P. (1986). Pressure to publish and fraud in science. *Annals of Internal Medicine, 104*(2), 254–256. https://doi.org/10.7326/0003-4819-104-2-254.

Wowra, S. A. (2007). Moral identities, social anxiety, and academic dishonesty among American College Students. *Ethics and Behavior, 17*(3), 303–321. https://doi.org/10.1080/10508420701519312.

Yang, S. C. (2012). Attitudes and behaviors related to academic dishonesty: A survey of Taiwanese graduate students. *Ethics & Behavior, 22*(3), 218–237. https://doi.org/10.1080/10508422.2012.672904.

Zheng, S., & Cheng, J. (2015). Academic ghostwriting and international students. *Young Scholars in Writing, 12*, 124–133. Retrieved at: https://repository.usfca.edu/cgi/viewcontent.cgi?article=1000&context=rl_stu

Zwagerman, S. (2008). The scarlet P: Plagiarism, and the rhetoric of academic integrity. *College Composition and Communication, 59*(4), 676–710. Retrieved March 27, 2020, from www.jstor.org/stable/20457030

References for Case Study: The Temptations of Experimental Social Psychology

Abma, R. (2013). *De publicatiefabriek. Over de betekenis van de affaire-Stapel* [The publication factory. On the significance of the Stapel affair]. Nijmegen: Vantilt.
Ellemers, N. (2013). Connecting the dots. Mobilizing theory to reveal the big picture in social psychology (and why we should do this). *European Journal of Social Psychology, 43*(1), 1–8. https://doi.org/10.1002/ejsp.1932.
Hull, D. L. (1998). Scientists behaving badly. *New York Review of Books, December, 3*, 24–30.
Levelt, P., et al. (2011). *Interim-rapportage inzake door prof. dr. D.A. Stapel gemaakte inbreuk op wetenschappelijke integriteit*. Tilburg: Universiteit van Tilburg.
Levelt, P., et al. (2012). *Flawed science. The fraudulent research practices of social psychologist Diederik Stapel*. Tilburg: Tilburg University.
Stapel, D. A. (2008). *Op zoek naar de ziel van de economie. Over het werkwoord hebben en het werkwoord zijn* [Searching for the soul of the economy. On the verb 'to have' and the verb 'to be']. Tilburg: Tilburg University Press.
Stapel, D. A. (2011, October 31). *Public statement*.
Van Lange, P. A. M., Buunk, A. P., Ellemers, N., & Wigboldus, D. H. J. (2012). *Sharpening scientific policy after Stapel*. ASPO: Internal report. https://kli.sites.uu.nl/wp-content/uploads/sites/426/2019/09/Sharpening-Scientific-Policy-After-Stapel.pdf

References for Case Study: Unexplained or Untrue?

Draaisma, D. (1970). Een Replica van Heymans' onderzoek betreffende telepathie [A replica of Heymans' study into telepathy]. *Heymans Bulletins*, p. 79.
Heymans, G., Brugmans, H. J. F. W., & Weinberg, A. A. (1920). Een experimenteel onderzoek betreffende telepathie [An experimental study regarding telepathy]. *Mededeelingen der SPR, 1*, 1), 1–1), 7.
Linschoten, J. (1959). Parapsychologie en algemene psychologie' [Parapsychology and general psychology]. *Tijdschrift voor Parapsychologie, 27*, 226–237.

Chapter 6
Falsifying

Contents

Electronic Supplementary Material: The online version of this chapter (https://doi.org/
10.1007/978-3-030-48415-6_6) contains supplementary material, which is available to authorized users.

© The Author(s) 2020   117
J. Bos, *Research Ethics for Students in the Social Sciences*,
https://doi.org/10.1007/978-3-030-48415-6_6

After Reading This Chapter, You Will:

- Know exactly what falsifying is
- Be able to distinguish falsification from other forms of fraud
- Understand how falsifying impacts the social sciences
- Develop strategies to address falsification

Keywords Conformation bias · Deep data diving · Dichotomization · Disconfirmation dilemma · Editorial bias · False positive · Falsification · Falsifying · File drawer problem · Image manipulation · Impact factor · HARKing · My side bias · Outliers · Pathological science · Peer review · p-hacking · Post-publication · Protheus phenomenon · Publication ethics · Publication bias · Questionable research practice · Replication · Retraction · Reviewer bias · Self-correction · Self-deception · Sloppy science · Submission bias

6.1 Introduction

6.1.1 Sloppy Science

In 2012, Anthropologist Mart Bax of Free University Amsterdam was already in retirement when suspicion arose about the validity of his field work, some of which dated back to the 1970s and 80s. It was thought that if he hadn't fabricated his research outright, then at the very least he had manipulated it. That is to say, he was accused of (among other things) having altered and even removed crucial details in his data. He had ascribed statements to untraceable participants and staged actions that could not be verified.

An integrity commission was set to work and concluded the following year that Bax was guilty of scientific misconduct. He had presented 'improbable events as "historical facts," embedded in research that systematically obscures names of persons and places and muzzles sources and contains inaccuracies in a large number of places' (Baud et al. 2013, p. 39).

The lack of openness and transparency Bax exhibited has become an exemplar of what we now call 'sloppy science' – carefree and negligent research practices that include both intended and unintended violations of scientific norms. In this case, the researcher seemed to have placed little value into verifiability and transparency, which sit as cornerstones of appropriate scientific practice.

When sloppy research veers into falsehoods, we speak of 'falsifying' or 'falsification.' To avoid confusion with the terminology used by philosopher Karl Popper (see below), we stick to 'falsifying.' Falsification in Popper's sense means actively seeking to *dis*confirm a hypothesis, falsifying effectively amounts to the opposite.

6.1.2 Falsifying

The term falsifying, literally meaning 'rendering false,' entails forms of manipulation that allow researchers to use a dataset that supports biased or even erroneous claims. It includes 'trimming' (leaving out certain findings) and 'massaging' (slightly changing) data, as well as altering images, misrepresenting results, and simply *not* reporting findings.

If fabrication (presenting fake or non-existent research data) and plagiarism (literary theft) are sciences' deadly sins, punishable by severe penalties, then falsifying is its *daily* sin. It is a less visible, less spectacular form of misconduct, and the scientific community tends to view it more tolerably than its lethal counterparts. Unjustly so, says Köbben (2012), who warns that the accumulation of these smaller sins represents a far greater danger to science than the (isolated) larger ones. Overtime, and left unchecked, this accumulation will result in the large-scale pollution of scientific research.

This being said, some considerations must be first addressed. Not all manipulations represent a researcher's intent to deceive, nor is every form of manipulation prohibited, as we shall see below. Also, one must distinguish between deliberate manipulations and honest errors. Furthermore, these two should not be confused with scientific disagreement (researchers challenging the conclusions of one another). Thus, although falsifying is considered 'misconduct,' it can be difficult to assess exactly *when* acceptable research practices lapse into dubious ones. It is on this note we enter into the heart of the problem of academic fraud – which is less about demarcating right from wrong, and more about ethical reflection and decision-making. The aim of this chapter is to raise awareness of these issues by exploring several dimensions of falsifying in research practices.

By and large, in this chapter we follow the research process itself. We start with the forms of bias that appear at the first stage of the process, when research questions are posed. Following the selection of questions, a discussion will take place regarding the falsities that result from (slight) alterations, or the act of manipulation during data collection and analysis.

We finish with a discussion of the biases often present when research conclusions are reported and disseminated in a skewed or one-sided way. Though this is referred to as *publication ethics,* it relates to our subject of research ethics because it discloses disturbances in the research process.

In two separate sections, we discuss the problem of *self-deception* (falsifying by not being critical enough) and a possible remedy against the falsehoods inherent to science.

6.2 Bias at the Start of the Research Process: Asking Critical Questions

6.2.1 Confirmation

Research starts with asking questions. But, as any student knows, asking good questions demands a self-critical attitude, and a readiness to address and counter one's own preconceptions. In fact, scientists should actively look for information that disconfirms their opinions about the world, an action for which the term 'falsification' is reserved (as explained above).

The reality is, this is much more difficult than it appears. There is a long-identified experimental effect known as *confirmation bias* (sometimes called *myside bias,* see; Perkins 1985; Toplak and Stanovich 2003). Confirmation bias consists of the tendency of individuals to judge new information in a way consistent with their preexisting ideas or convictions. People thus prefer supporting information rather than conflicting information and tend to overlook or disregard information that does not fit into their worldview (Jonas et al. 2001).

Notorious examples are found in 'psychic studies' (studies into paranormal activities) and psychoanalysis (studies into the unconscious mind). In both traditions, there exists a strong tendency toward confirming what was theoretically hypothesized. But it is far from restricted to just these domains and has been observed in more empirically oriented research traditions a well.

As a case in point, Greenwald et al. (1986) examined empirical research into a phenomenon known as the *sleeper effect.* This is the counter-intuitive finding that a persuasive message accompanied by a 'discounting cue' (a prompt that indicates the message is untrustworthy) tends to develop more impact over time. For example, viewers watching a 'smear campaign' against one political candidate, paid for by the opposite candidate, will develop a more favorable attitude towards the message weeks *after* being exposed to the message, rather than immediately afterwards, despite being aware that the source is biased.

To explain this effect, it was hypothesized that over time, the discounting cue becomes dissociated from the original message, and therefore ceases to be effective in countering it (this is known as the *dissociation hypothesis*). Research into the sleeper effect has not been able to confirm this hypothesis, however that did not deter the researchers from investigating it. Only much later did researchers realize that the 'dissociation hypothesis' was incorrect, and that an entirely different explanation was required. This fixation on a single hypothesis, and the resulting neglect of alternative theories has obstructed scientific understanding on the subject for some 25 years, Greenwald et al. observed.

Box 6.1: 'Disconfirmation Dilemma'
How are we to deal with disconfirmation in the empirical process? If the results of an experiment don't confirm theoretical expectations, researchers are confronted with what Greenwald et al. call a 'disconfirmation dilemma.' Researchers can decide to either (a) reanalyze the data, (b) revise the procedures, or (c) reformulate a different prediction (based on the same theory). Rarely do researchers decide for option (d), to publish disconfirming results.

When researchers resolve the disconfirmation dilemma by repeatedly retesting predictions, instead of reporting disconfirmation, they may be accused of some form of 'falsifying' because they resort to theory-confirmation rather than theory testing (Fig. 6.1).

6.2.2   Challenging Bias

Effectively, confirmation bias undermines open and critical thinking and runs against creativity. This poses a serious challenge for science. Open-mindedness and creativity are two of science's most crucial features. How can scientists avoid or

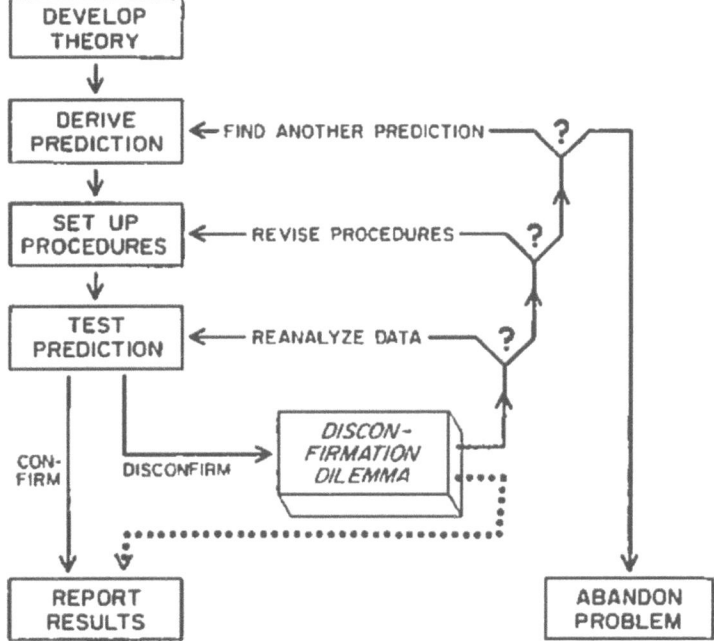

Fig. 6.1 Disconfirmation Dilemma. (Adopted from Greenwald et al. 1986, p. 220. The dotted line represents a route infrequently taken)

counter this type of bias? Is it possible, for example, to train scientists to consider *both* sides of an argument? There is evidence that this may be possible, at least to a degree.

Wolfe and Britt (2008) found that when students were assigned to one side of a (somewhat controversial) topic and received instructions to search for as much information on the topic as they saw fit, they would display confirmation bias. However, when instructed to search specifically for balanced information, confirmation bias was significantly reduced. Similarly, Macpherson and Stanovich (2007) found that *decontextualization instructions* (instructions to put aside one's own convictions and consider the issue from opposite sides) helped reduce confirmation bias. These findings, preliminary as they are, point to the importance of making explicit one's expectations (Box 6.1).

6.3   Bending the Empirical Cycle: Manipulations During Research

6.3.1   'Lies, Damned Lies, and Statistics'

The next step in the research process consists of setting up a design in order to test hypotheses. In the social sciences, significance testing of null hypotheses is an omnipresent tool, and will be the focus of the next three sections.

The simplest example of significance testing is the evaluation of null hypothesis $\mu_E = \mu_C$ against the alternative hypothesis $\mu_E \neq \mu_C$. Here μ_E denotes the mean of the outcome variable in an experimental group, and μ_C the mean of the outcome variable in the control group. If the p-value for testing the two hypotheses against each other is smaller than .05, the null hypothesis is rejected. If it is larger than .05, then it means there is no significant difference between the two groups, and the null hypothesis is accepted on account that no evidence of an experimental effect was found.

Scientific journals have long tended to only publish results that have shown the experimental condition to be 'effective,' that is, if the p-value is smaller than .05. It is therefore crucial for researchers to obtain low p-values, otherwise their effort, time, and money is wasted. For some, obtaining small p-values has become a goal in itself. This raises some ethical questions which will be explored below.

6.3.2   Questionable Research Practices

Can research outcomes be manipulated such that lower p-values are obtained? The answer is yes, for example, by removing so-called 'outliers.' Outliers are extreme scores, and removing them heightens the chance of getting significant results. It sounds like cheating but that need not be the case, there can be good reason to remove outliers. Outliers can result from data errors (incorrectly recorded data) or because respondents may have failed to understand their role, or the questions asked. For instance, one survey gathered data on nurses' hourly wages. While on average respondents reported to earn $12.00 an hour, with a standard deviation of $2.00, one nurse reported an hourly wage of $42,000.00, which was clearly erroneous (it was more likely their annual income). Not removing this number would influence the true outcome (see Osborne and Overbay 2004).

However, consider the case of Dirk Smeesters, professor of consumer behavior and society at the Rotterdam School of Management. His work attracted the attention of fellow researcher Uri Simonsohn from Wharton University in 2011. He had read some of Smeesters' work and suspected foul play. Simonsohn believed Smeesters' studies were 'too clean to be the result of random sampling' (quoted in Chamber 2017, p. 81). He requested and obtained Smeesters' dataset and discovered anomalies. It seemed that Smeesters had removed participants from his data when they led his hypotheses toward not being confirmed. Smeesters responded that the participants 'had not understood the instructions' (quoted in Kolfschooten 2012, p. 270).

This did not satisfy Simonsohn. An integrity commission investigated the case and ruled that this reversal of logic, by which outliers are removed to boost significance, should be understood as 'data massaging.' Smeesters confirmed that he had acted 'erroneously' but denied that he had committed fraud: 'What I have done was to give a study, which was already almost good, a push in the right direction' (Kolfschooten 2012, p. 270). That didn't help his case. Seven papers he co-authored were retracted and Smeesters resigned from his position in 2012.

What Smeesters engaged in are called *questionable research practices*, or QRPs for short. QRPs have become serious concerns in the academic community. Simmons (2011) noted that 'flexibility in data collection and analysis allows researchers to present almost anything as significant.'

QRPs take many shapes and forms. To name a few; failing to report all dependent measures, selective reporting (only submitting studies that were successful), and excluding data after looking at the impact (as Smeesters had done).

Evidence of QRPs on a large-scale were found by Masicampo and Lalande (2012). They collected the reported p-values from three high-level psychological journals and compared their distribution. Given that smaller p-values are more appreciated, one would expect to see a steady decline in reports with larger p-values. What they found instead was a steady decline, followed by a peculiar peak of p-values just below .05 (see Fig. 6.2).

Many take this as evidence for the existence of 'falsifying,' because it appears that researchers have manipulated their data to ensure their data falls within an acceptable p-value of below .05.

Further qualitative evidence of QRPs on a large-scale was found by Leslie John and his collaborators. John et al. (2012) surveyed over two thousand psychologists and found a majority of psychologists admitted to engaging in a variety of such behaviors. In their widely circulated article, it was estimated that some questionable

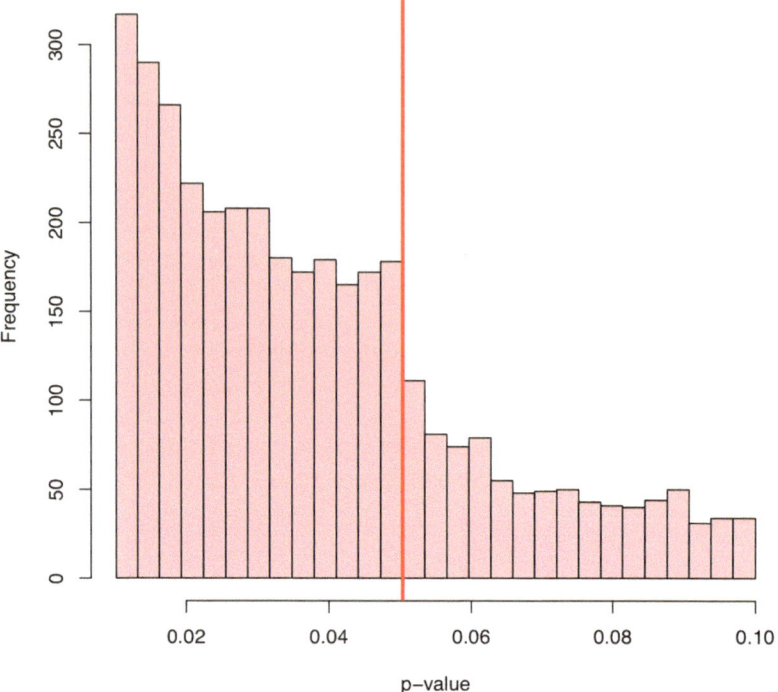

Fig. 6.2 'A peculiar prevalence of p-values just below .05'. Figure by Larry Wasserman, based on the data collected by Masicampo and Lalande (2012). Used with permission from the author. (Source: Normal Deviate, entry August 16, 2012)

research practices are so widespread that it must be assumed that virtually *everyone* uses them.

This raises the question as to whether these practices constitute a new scientific norm (John et al. 2012). Discussing this controversial conclusion, Fiedler and Schwarz (2016) warn against an inflation in the usage of the term QRPs precisely because of the suggestion of normalization. Some of the reported practices, they argue, are merely ambiguous, not 'questionable,' while others may or may not be justifiable depending on the specifics of the case (p. 50).

We do not propose that QRPs are the 'new norm.' On the contrary, there is a serious danger that the scientific literature becomes polluted with 'breakthroughs' (significant findings) which are not breakthroughs at all. Indeed, in the June 1st, 2011 issue of *Scientific American,* John Ioannidis argues that exaggerated results in peer-reviewed scientific studies have reached 'epidemic proportions' in recent years (Box 6.2).

Box 6.2: 'P-hacking and HARKing'

The following case, discussed in several entries on the critical website *Retraction Watch* in 2017, provides a rare glimpse into how the academic community responds to questionable research practices.

Several years ago Brian Wansink, a world-renowned food researcher at Cornell University, provided a visiting PhD student with the complete set of data of a self-funded study that failed to produce any notable results. He told the student that it is was well worth the effort to search for overlooked patterns, further stating that 'there's got to be something here we can salvage because it's a cool (rich & unique) data set.' The student set to work and managed to produce five articles in just six months using the dataset. In a now deleted blog post of November 2016 ('The Grad Student Who Never Said "No"'), published on his personal website, Wansink proudly reported on this student's success, presenting it as a 'lesson in productivity'.

His readers were less impressed. 'This is a great piece that perfectly sums up the perverse incentives that create bad science. I'd eat my hat if any of those findings could be reproduced in preregistered replication studies.' Another reader commented, saying that what was described in the blog sounded suspiciously like p-hacking and HARKing (entry at *Retraction Watch*, 2.2.2017).

P-hacking (also called 'phishing') is term used to describe how researchers try to uncover statistically significant patterns in a data set without having a specific hypothesis. They just hope to find statistically significant results. *HARKing* is the flipside of this coin (HARK stands for Hypothesizing After Results are Known). It consists of presenting a post hoc hypothesis in a research report as if it were, in fact, an 'a priori' (earlier formulated) hypothesis.

Had Wansink been 'bending the rules of the game' by letting his student go through raw data in the hopes of unearthing something (*anything*), which then would be presented as a 'finding'?

When confronted with the accusation of p-hacking, Wansink retorted that testing the null hypothesis had been his 'plan A.' It was when he didn't find anything that he turned to 'plan B.' As Wansink explained: 'P-hacking shouldn't be confused with deep data dives – with figuring out why our results don't look as perfect as we want. With field studies, hypotheses usually don't "come out" on the first data run. But instead of dropping the study, a person contributes more to science by figuring out when the hypo worked and when it didn't. This is Plan B' (quoted in an entry on *Retraction Watch*, 2.2.2017).

Wansink's rebuttal failed to convince readers of *Retraction Watch*. One wrote: 'Deep dives are great, but they should be planned when the study is being constructed, not created after the fact in an attempt to "salvage" something from the experience' (2.2.2017). Another sarcastically remarked: 'Wansink's use of the phrases "our results don't look as perfect as we want" […] pretty much speaks for itself' (3.2.2017).

(continued)

Box 6.2 (continued)

However, not everyone saw wrongdoing. Another reader wrote in Wansink's defense, exclaiming that not all exploratory studies constitute 'p-hacking': '[There] is nothing wrong with a researcher honestly engaging in and presenting an exploratory analysis without a single pre-defined hypothesis. As long as these studies are presented honestly, they can provide useful insights and generate useful hypotheses that can later be verified (or debunked) through attempts to replicate, often by other researchers in the field' (19.2.2017).

The result of Wansink's actions? – By the summer of 2019, when this chapter was written, 17 of his papers were retracted (one even twice) and he resigned from his position. (See Resnick and Bellus 2018, for further discussion.) (Fig. 6.3)

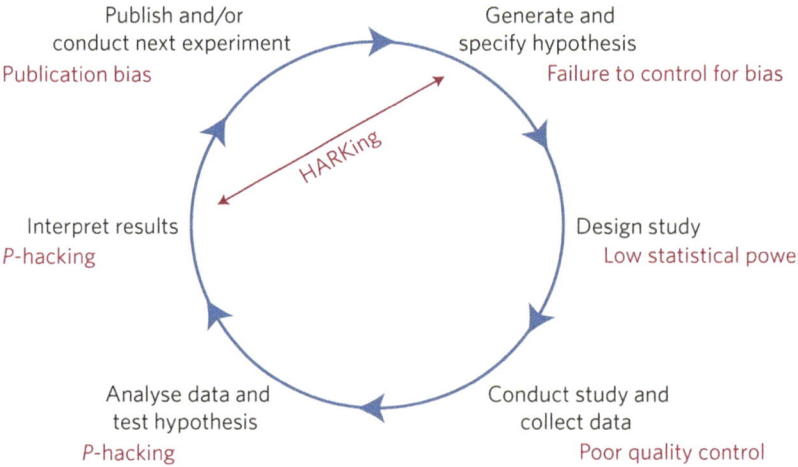

Fig. 6.3 Threats to reproducible science. (Adapted from Munafò et al. 2017)

6.3.3   *Image Manipulation*

Recent technological advances provide researchers with a wealth of opportunities for furthering their research in ways not available fifteen or twenty years ago. But as with any change, new ethical considerations emerge. In the case of digital images, there is a growing concern in the scientific community over how to properly handle them.

Today, there are multiple known cases of unethical manipulation of images that affected the interpretation of the data presented, a number of them having led to retractions. One such case, a 2009 paper published in the *Journal of Biological Chemistry* by Spanish researcher José G. Castaño, was retracted because (quoting from the notice published September 9th 2016 on *Retraction Watch*) 'the same image was used to represent results of different experimental conditions' on multiple occasions, adding further that the 'background of one image had inappropriately been adjusted.'

A lack of awareness of what is considered an acceptable form of image manipulation calls for the creation of guidelines to help researchers distinguish between appropriate and inappropriate use of digital images. A set of such general principles is discussed by Cromey (2010), who compares these guidelines to what is already established practice in the field of photojournalism. A sampling of these guiding principles includes the following recommendations:

- Digital images should be acquired in a manner that does not intend to deceive the viewer or to obscure important information.
- Manipulation of digital images should be performed only on a copy of the image.
- Simple adjustments and cropping are acceptable but lossy (irreversible) compression should be avoided, and use of software filters to improve image quality is not recommended.
- Cloning or copying objects into a digital image is considered highly questionable (Box 6.3).

Box 6.3: 'Consequences of Retraction'
With an increased awareness of research ethics in our day and age comes an increased awareness of the consequences of misconduct. We quote from an anonymous cry for help, posted on October 16th, 2013 on 'Editage Insights' (a platform for researchers, authors, publishers and academic societies): 'I recently got an email from the editor of a journal in which my paper is published, requesting me to retract the paper because they found some errors in my data and statistical analysis. I am worried about my reputation if I have a retracted paper. I may not get a grant for my next study. Please advise me.'

In a response posted on March 30th, 2017: 'I would encourage you to respond positively to the journal editor's request and offer to have your paper retracted. If you do so, the journal's retraction notice will inform readers that the paper has been retracted by agreement among the authors and the journal editor, owing to errors in data.' (source: Editage Q&A).

6.4   Bias in Disseminating Research: Publication Bias

6.4.1   File-Drawer Problem and False Positives

Imagine a researcher testing the effects of the new and promising treatment 'X' (say a particular form of cognitive behavioral therapy for a certain type of anxiety disorder). To the researcher's disappointment, a comparison between the experimental group who received this form of therapy and a control group who received no therapy (or a different therapy), resulted in a p-value of .14. Since this is larger than .05, the hypothesis is rejected, and the results are not published. Another researcher (unaware of the first researcher's work because it was not published) is interested in the same therapy. Their comparison results in a p-value of .37, and again the results are not published. Shortly thereafter, a third researcher evaluates treatment 'X'. They find a p-value of .02, and as a result the findings are published.

Based on this one publication by researcher No. 3, the unsuspecting reader will conclude there is evidence for the effectiveness of treatment 'X'. In reality, the effectiveness of treatment 'X' is a *false positive* because there is actually more evidence to the contrary – that it doesn't work. The problem being, that evidence was never published. Our full understanding is obscured by the fact that studies that show no result are not published. Robert Rosenthal coined a term for this; the *file-drawer problem*.

'False positives' and the complementary 'file-drawer problem' relate not so much to theoretical or methodological issues in research, but to questions regarding *dissemination* (communication or non-communication of research findings), related to *publication ethics*.

If there are frequent false positives and/or numerous unpublished null results, any meta-analysis of a particular research subject will turn up corrupted. How serious is this problem?

One way of researching this question is by comparing the amount of research undertaken with the number of publications stemming from that research. In many fields of research, pre-registration is required (the researcher must catalogue his research protocol in advance, submitting hypotheses, methodology, and expected findings). This makes it easier to check for both p-hacking and HARKing, but it also allows for the questioning of *submission bias* (the tendency to only submit for publication studies that have 'positive findings'). In the social sciences though, pre-registration is a recent phenomenon and not the norm, with a few exceptions.

One exception is the public registry TESS (Time-Sharing Experiments in the Social Science). Franco et al. (2014) followed studies registered in TESS over a ten-year period, to see how many of them were eventually published in peer-reviewed journals. It turned out that 80% of the registered studies were written up, but less than half (48%) were published. Unsurprisingly, there proved to be a strong relationship between the outcome of the study (whether or not the hypothesis was supported by the results) and it being published. Studies that had negative results were far less likely to be published, and even less likely to be written up at all.

Why do researchers opt to not write up 'null results'? Franco et al. (2014) questioned a selective group of researchers by email and got answers that confirmed their suspicion. As one of their respondents reported: 'I think this is an interesting null finding, but given the discipline's strong preference for $p < 0.05$, I haven't moved forward with it.' (p. 1504).

6.4.2   Reviewer and Editorial Bias

Another source of publication bias is located in the process of peer review and journal editorship. Both peer reviewers and journal editors effectively function as gatekeepers, deciding whether an article is worthy of publication. We discuss both roles below, starting with peer reviewers.

The 'peers' in the peer review process are researchers themselves, often experts in a field from which they are recruited. They are asked to assess the quality of manuscripts sent to a journal. The review procedure in which they participate is as a rule blind. That is to say, the author is not aware of the reviewer's identity (single blind), but often the reviewer doesn't know the identity of the author either (double blind). Peer reviewers do not get paid for their work, should have no interests involved, be unconnected to the authors, and act solely on the desire to guarantee objectivity and impartiality in science.

But does it really work like that? Some argue that peer review is indeed the best system that we have, providing impartial quality control. Others contend it may have functioned as such at one point but it longer does in today's society, where science cannot permit the luxury of operating from an ivory tower any longer (we turn to this discussion in greater detail in Chap. 9). And then there are those who argue that the peer review system has never guaranteed quality control in the first place.

They point to the fact that some of the most important and groundbreaking works in the history of science were never peer reviewed, that some of these work were initially rejected by peer reviewers, and that vice versa, flawed, non-sensical or even absurd papers were accepted by them (see Box 5.7 on 'hoaxing').

If we cannot rely on peer reviewers to detect errors, identify misconduct, or spot what is truly innovative, then the peer review system does not safeguard quality. But perhaps it is even worse. There is reason to believe that the peer review system is biased in at least two ways.

1. Peer reviewers may be too conservative. Reviewers are believed to be biased against new findings and new ideas. Also, they focus too much on finding weaknesses in manuscripts and not on the positive contributions therein. Suls and Martin (2009, p. 43) argue this may be so because 'appearing to be too lenient seems worse that appearing to be too harsh.'
2. Reviewers seem prejudiced in favor of prestigious research institutions and established authors. Peters and Ceci (1982) found confirmation of this suspicion in a small-scale study they performed. They selected 12 previously published

articles, originally written by researchers from prestigious American psychology departments, and resubmitted them to the same (highly prestigious) journals that had previously published them, but under fictitious names and fictitious institutions. Only three articles were detected as 'resubmissions'; eight out of the nine remaining articles were rejected on ground of insufficient quality (some were even critiqued for having 'serious methodological flaws').

Consider next the role of journal editors. As gatekeepers, editors not only have to safeguard quality, but also to present interesting, original, and novel findings to their readership. This may lead to a bias against replication studies because they don't offer anything new, despite replication being the 'gold standard' in science (see Møller and Jennions 2001). Kerr, and later Rowney and Zenisek (quoted in Hubbard and Armstrong 1994) conducted a survey among editors and review board members of both management and psychology journals, and indeed found confirmation of *editorial bias* against replication studies.

The editor's obligation to present novel and interesting results can furthermore lead to an effect known as the *proteus phenomenon*. In essence, whenever positive results are published, a window of opportunity quickly opens for researchers to publish findings that contradict these results. Editors often publish these findings because they too are a 'novelty.' The net effect is a tendency for journals to rapidly publish conflicting results (see Pfeiffer et al. 2011) (Box 6.4).

Box 6.4: 'Adjusting the Data? A Dilemma'
On the 'r/*AskAcademia*' Reddit community, a student identified as 'Throwinbin' (henceforth 'T') published a telling post. T referenced a supervisor who requested that they make use of a dataset in which a specific approach was implied. The problem was, the data (provided by a third party) did not fit the format that the approach required. In T's words: 'Using [my supervisor's] method will involve removing whole articles from our data set and changing the (important, central) main attribute of the set. It's basically massaging the data until it fits the model of his method. I'm not comfortable doing this as I strongly believe that it's going to give us false results […]. How to raise this with [my supervisor] without sounding really bad?'

Below are the (edited) exchanges between T and three community members who responded to the post. From these exchanges, it becomes clear that 'data massaging' is not the only issue at stake, and a number of other ethical dimensions are in play. Consider how T managed the situation:

Respondent A: 'Is there a person in your department you could consult (with no stake in your publications)?'

T: 'I have a "second supervisor," but I'm not keen to take this concern elsewhere in the department at the moment. I like my supervisor and I don't want to harm his career.'

(continued)

Box 6.4 (continued)

Respondent B: 'If you do want to approach your supervisor, then ask how to organize the data to make them fit (rather than the approach of saying it is wrong), and maybe he will explain something you hadn't thought about.'

T: 'I've done this – which is how I now have in emails him telling me to remove certain rows from the data, and later reorganize whole columns without making sure the changes carry over the entire dataset. I've made the changes he asked for and ran his method and it looks pants [not good].'

Respondent C: 'Could you not go down the 'play dumb' route? – "I'm confused, maybe I'm just stupid, I'm not sure how your [method] is entirely relevant. Can you explain how it's better than x"?'

T: 'I do understand his need for this method well; he wants to move on from our current institution and this will look good on a CV. I sympathize with him as I'm not very happy with the situation in our department either and would be looking to move on if I could. I've admitted defeat and made our data work his model. Results so far are rubbish, so I'm going to take it to him and put the ball in his court – though if he insists on it going into a paper I don't want my name anywhere near it.'

You: What advice would you give 'T'?

6.5 Self-Deception

6.5.1 Mistakes Happen

Earlier we noted that various forms of falsifying must not be confused with 'honest mistakes.' But there is one type of 'honest mistake' that *should* be considered a form of falsifying, even if the researcher had no intention to mislead. This is *self-deception*. Self-deception occurs when the researcher is so strongly convinced that a particular model or theory is correct that they are unable to accept evidence to the contrary. We will discuss a few forms of self-deception from here.

Perhaps the strongest form of self-deception consists of discovering information that doesn't exist, a phenomenon ironically dubbed *pathological science*. The discovery of so called 'N-rays' (an alternative to X-rays) by French physicist René-Prosper Blondlot in 1903 counts as one such example (Grant 2008, pp. 88–89). Blondlot built a device that enabled him to 'see' these rays. He gave demonstrations and others, if trained properly (or told what to look for), would see them too. The non-existence of N-ray was exposed when a sceptic secretly turned off the device and Blondlot still claimed to 'see' the rays.

The biography of Wilhelm Reich, a former Freudian who had gone astray (Sharaf 1983) offers a similar story. Reich was convinced he had discovered a new form of energy, which he called 'orgone.' He built a device in which orgone would accumulate, and while testing his device, he found a constant temperature difference of 2 °C inside the 'orgone accumulator.' Believing this to prove the validity of his discovery beyond a reasonable doubt, he contacted Einstein, who kindly agreed to study his device. Two weeks later, Reich received word from Einstein, who stated that his assistant had come up with a simpler explanation for the temperature difference – lack of air circulation. Reich was unswayed and maintained faith in his 'discovery.'

In the field of parapsychology, Alfred Russell Wallace, a British naturalist and the co-discoverer of natural selection, offers another interesting example. During his investigations, Wallace accepted certain observations as evidence for the existence of extra-perceptual phenomena. In his autobiography, he wrote about his conversion to 'Spiritism' after having attended a series of séances with a 'medium': 'I was so thorough and confirmed a materialist that I could not at that time find a place in my mind for the conception of spiritual existence, or for any other agencies in the universe than matter and force. Facts, however, are stubborn things. […] Facts became more and more assured, more and more varied, more and more removed from anything that modern science taught, or modern philosophy speculated on. The facts beat me' (quoted in Shemer 2002, p. 192).

The irony of 'facts' having 'beaten' Wallace is probably not lost on the reader, for the séances were in reality very likely carefully orchestrated performances of frauds. Wallace himself, however, was not a fraud – he was taken in by the performance (Fig. 6.4).

Fig. 6.4 Carried away by self-deception

6.5.2 To Remember or Not to Remember (That Is the Question)

The above discussed cases of self-deception may be comical illustration of how scientists were able to fool themselves in the past. Yet the question arises whether we can be sure that some of our own present day discoveries are not also instances of self-deception.

In the last two decades of the twentieth century, a debate emerged within psychology over whether or not a repressed childhood memory of sexual assault could be recovered through the aid of specific forms of therapy (Pezdek and Banks 1996). Proponents of recovered memory therapy argue that children who go through such assaults 'dissociate'; meaning they repress all memories of such traumatic experiences and will not remember them, unless aided in some way.

Recovered memory therapy became prominent in the 1990s. Certain recipients of the therapy recalled highly bizarre satanic-abuse memories. In some cases, these testimonies led to prison sentences for men accused of these crimes. However, it turned out that at least some of these accusations were false and the 'perpetrators' were released. This prompted critics to question the validity of recalled memories (for a full discussion see Loftus and Ketcham 1994).

Did the 'recovered memory movement' find in the testimonies of their clients what they *wanted* to hear, or had they unearthed a new phenomenon which mainstream science *refused to accept* because it was too controversial? (Box 6.5)

Box 6.5: 'Not Sure If It's Research Misconduct'
A PhD student turned February 2020 to 'r/AskAcademia' discussion platform on the website *Reddit*, for advice, writing how that it seemed as if a professor in had been engaged in research misconduct, and considered taking it to the board.

When I opened the file [of my professor], I noticed that a lot of the figures have been altered. In several cases he took a bar graph (looks like a screenshot from a Prism file) and then covered one of the bars with a different bar. The new bar would have a different height and number of significance asterisks than the original one. I could move over the replacement bar and see the original figure – the replacement was clearly cropped out of a different figure and pasted onto this one. [...] I talked to a few other students about this. They think it's possible he's just lazy or doesn't know how to use Prism. Like maybe his students repeated an experiment and that changed the results, but he didn't want to (or know how to) update the figure in Prism so he just pasted the new bar on top? This seems sketchy to me, and I don't think it explains the difference from his published figure either. [...] I'm hesitant to ask him directly. If he actually is falsifying data, it's not like he's going to admit it to me. I'd prefer to speak with the department chair and see what she recommends.

(continued)

Box 6.5 (continued)
However my classmates think going over his head without first asking for an
explanation would be wrong. I'm really at a loss for what to do.
   Here are some of the replies this PhD student received:

1. I think it's quite weird that you are worried that it may reflect badly on you
   if you ask him directly. Yet, you think that the more drastic approach of
   going to the chair is less worrying.
2. You don't have evidence of misconduct and there are only downsides to
   yourself from making accusations. I'd say forget about it.
3. I think a good and non-accusatory way to go about it is (if initially via
   email): 'Hi X, I've noticed that there are revisions to the graphs in the
   PowerPoint. Did you happen to obtain more evidence/data changing the
   original graphs and supporting your conclusions? If so, what steps or projects are you pursuing after the new information?'

   Which of these advices do you prefer? Or would you consider a different
approach?
   Source: Reddit, AskAcademia, 'Not Sure if What I'm Seeing is Research
Misconduct?'

6.6   Science's Self-Correction

6.6.1   Self-Correction

Discoveries of falsehoods in research are traditionally met with a defensive system
of self-correction known as *retractions*. This quite simply means that a 'contaminated' publication is flagged but not withdrawn from the public domain. A note is
attached to the paper that states it has been 'retracted.' Retraction can take place
with or without the author's consent and can be argued on the grounds of methodological or theoretical flaws, or because research misconduct was identified.
   Retractions are not to be taken lightly. Until quite recently, misconduct and subsequent retraction of a publication remained an internal matter, known to only a few
parties. However, with the rise of digital publications, retractions have become
much more public (and visible). For example, the website *Retraction Watch* is dedicated exclusively to highlighting misconduct, fraud, and retractions in science
(across all disciplines). It keeps track of virtually all that is going on in the academic
world in a very public manner, posting the full names and affiliations of all parties
involved. A retracted article, though 'withdrawn,' not only remains visible, it effectively becomes a permanent stain on an author's reputation (see Box 6.3 for an
impression of this consequence).

Apart from the personal consequences, there remains the question of what damage fraudulent articles can cause. After all, an undetected (unflagged) fraudulent paper will remain in the public domain, continuing to act as a source of pollution in future literature. This is important to consider, as a great deal of time may pass before a fraudulent study is retracted. Interestingly, in the last few decades, the retraction process has quickened, with the number of retracted papers increasing in lock step. From this, three questions can be raised: (1) Is this increase a good thing or not, and how to account for it?; (2) Are retractions the right answer to the problem of QRPs?; and (3) Are there better alternatives? We will very briefly touch on these issues in the sections to come.

6.6.2   Beyond Retraction?

In an often-cited article, Daniele Fanelli (2013) investigates retractions in scientific literature. Scouring through data from the Web of Science (a publisher-independent global citation database) for the entire twentieth century, Fanelli notes a sudden increase in retracted papers per year since the 1980s of some 20%. He then compared this increase to the number of 'corrections' applied to articles in the same period, which did not see a similar increase.

Fanelli proposed two hypotheses, that both have a radically different outlook on the question of whether or not the increase in retractions signifies a positive development. One attaches the increase to growing misconduct within the academic community, and thus sees the growing number of retractions as a bad sign. The other states that the system has become more resilient, and thus the increased number of retractions signifies something good.

Fanelli argues that the evidence in his study suggests that the 'stronger system hypothesis' is more likely to explain the rise in retractions than the hypothesis that scientists have become more fraudulent. Peers, editors, and the scientific community at large seem to have become more sensitive to and aware of misconduct, and consequently, have become more proactive about it (see furthermore Ioannidis 2012; Fanelli 2018).

This begs the question, even if editors have become more aware of the issue, can we trust that science will be able to rectify (all of) its mistakes this way? Stroebe et al., reviewing a number of recent examples of misconduct, are not overly optimistic. Science is based on trust, they argue, and as such 'scientists do not expect their colleagues to falsify their data, and do not look for signs of fraud' (2012, p. 680). What would really help, Stroebe et al. argue, is to fortify the position of the whistleblowers, who, after all, have been responsible for detecting the majority of falsities in the first place.

Furthering this line of thinking, consider *Post-publication Peer Review* (PPPR). PPPR is a commenting system that allows publications to be reviewed and discussed online, on platforms such as *PubPeer* and *Open Review* after they have been published – on a (mostly) permanent basis.

Appraising this approach, Jaime Teixeira da Silva (2015, p. 37) considers that the advantage of PPPR is that it 'makes authors, editors, peers, journals and publishers accountable for what they have published or approved of publishing in the framework of their publishing models.'

However, the question is whether PPPR should consist of anonymous reviews (comparable with traditional peer review) or not. Teixeira da Silva is a vocal opponent of anonymity in peer review and a severe critic of *PubPeer,* which publishes anonymous reviews and allows unchecked accusations with little or no accountability.

Evidently, PPPR invites questions about the quality of those peers, but it also points to a new direction science is taking. In the twenty-first century, research in the social sciences is no longer considered an isolated effort of one individual (or a small group of individuals), but rather that of whole networks. Using the strength of collectives (networks) while simultaneously answering the increasing call for greater transparency, we find a growing inclination among social scientists to use open repositories to deposit and share data, pre-registration of protocols, and the commissioning of experts to monitor and review research. Thus, in the social sciences (modeling the medical sciences), ethical review boards have attained a progressively more important function in research.

While many of these initiatives further the social sciences in becoming more open and more accountable, aiding it in diminishing publication bias and forms of sloppy science, it does little to overcome confirmation bias, which still looms over the field, mainly because scientists will still only publish 'significant' results. In an attempt to address this problem, Ioannidis (2012) and van Assen et al. (2014), among other advocates, propose that journals should no longer focus on novel findings. Let them instead publish *everything*, including null results. They argue this change will make the scientific record complete, rather than fragmented.

6.7   Conclusions

6.7.1   Summary

In this chapter, we've followed the empirical cycle from beginning to end, exploring the various ways bias may disturb or corrupt our findings. We found that research does not always reveal what was intended or desired, leading to the danger of misrepresentation, one-sidedness, or even the production of downright falsehoods.

We showcased how the questions we ask may be biased towards the confirmation of what we already know. *Confirmation bias* (AKA myside bias) effectively obstructs creativity and progress in science and impedes more objective or at least impartial explorations from taking place.

With a strong incentive to publish research that show significant results, the danger of *questionable research practices* (QRPs) was introduced. *Data massaging, p-hacking, HARKing*, and other tricks meant to lower the p-value and thus 'heighten' the validity of research outcomes have the potential of polluting research findings on a large-scale, and endangers science's credibility.

The *file-drawer problem* and *false positives* point to the danger of bias during the dissemination process and reside under *publication ethics*. The tendency to report only what is significant, and to avoid reporting null findings creates a distorted view of reality, further enhanced by *editorial* and *review bias*, and the dangers of *self-deception*.

Increased retractions of 'contaminated' (fraudulent) papers show that science is able to self-correct, but the question is whether this is enough. Some argue that the system is strong enough to correct itself in the long run, whereas others believe more drastic measures are called for, including *post-publication peer review* (PPPR), *pre-registration,* and new journal policies to publish everything instead of only 'interesting', 'novel', and 'significant' findings.

6.7.2   Discussion

Research falsifying clearly poses a threat to science's claims of objectivity, verifiability, and other core values of science (see Chap. 2). Part of the problem may be attributed to overly ambitious researchers not taking the standards seriously enough, but part of it cannot be attributed to willful misconduct. Confirmation bias may be the result of something that remains entirely unconscious, and the file-drawer problem may be more likely the result of a fault in the system than the fault of an individual researcher. Similarly, editorial bias seems ingrained in the larger dissemination process, and certainly requires further attention. What suggestions do you have for addressing these ever-present issues of falsifying?

Case Study: Yanomami Violence and the Ethics of Anthropological Practices

Toon van Meijl

The classic monograph by Napoleon Chagnon ([1968] 1983) about violence and warfare among the Yanomami Indians in the Amazon is one of the best-known ethnographic studies of a tribal society. In combination with a number of films he made alongside Timothy Ash, the monograph *Yanomamo: The Fierce People* offered a penetrating picture into a society that was intensely competitive and violent. In Yanomami society, one-third of all adult men were claimed to die as a result of violence (Fig. 6.5).

Fig. 6.5 Yanomami
woman and her child at
Homoxi, Brazil, June
1997. (Photo: Cmacauley)

Chagnon visited the Yanomami periodically over many years to examine his assumption that patterns of warfare and violence may best be explained in terms of man's inherent drive to have as many offspring as possible, which he labelled reproductive fitness. He argued that the most aggressive men win the most wives and have the most children, thus passing their aggressive genes on to future generations more abundantly than the peaceful genes of their nonaggressive rivals. For Chagnon, the Yanomami provided an excellent case of this sociobiological principle because in the 1960s, while they were exhibiting an intense competition for wives, they were still virtually unaffected by Western colonial expansion.

The assumption that Yanomami society had not been influenced by colonial contacts, however, has been criticized by the historical anthropologist R. Brian Ferguson (1995). Rather than viewing the Yanomami as innately violent, he interpreted the intense violence in the region as a direct consequence of changing relationships with the outside world. Although the villages visited by Chagnon may not have had contact with missionaries or colonial officers, their presence in the wider region had disturbed the balance in inter-communal relations, especially by the introduction of steel tools and weapons. As a consequence, the rivalries between villages intensified and fighting erupted in efforts to gain access to the increasingly important new goods available in the region. Accordingly, Ferguson contended that the fighting was a direct result of colonial circumstances rather than biological drivers.

Several years later, Chagnon's interpretation of violence in Yanomami society was also criticized on ethical grounds by the investigative journalist Patrick Tierney (2000b). He argued that the violence witnessed by Chagnon had not only been caused by indirect influences of colonial contact with westerners, as Ferguson had argued, but also by Chagnon's own fieldwork practices. He pointed out in great detail that Chagnon had contributed to disturbing the balance between communities by providing steel goods, including weapons, to his informants, which in turn provoked numerous conflicts, raids, and wars. He was also accused of exploiting hostilities between factions and rival communities so he could document violent incidents for the films he and Timothy Ash produced. Finally, Chagnon was charged with transgressing Yanomami ethics by obtaining the names of dead relatives, which was considered taboo for surviving relatives. Thus, Chagnon's own fieldwork practices were argued to be a direct cause of the violence that he explained only in terms of genetics.

The publication of *Darkness in El Dorado* (Tierney 2000b) was preceded by a pre-publication in *The New Yorker* (Tierney 2000a), which appeared shortly before the annual meeting of the American Anthropological Association (AAA) in 2000. This piece highlighted an additional accusation, namely that Chagnon had collaborated with epidemiologist James Neel, who was claimed to have tested a new vaccine against measles among the Yanomami. As a consequence of Neel's work, hundreds of Yanomamis were said to have died because they never built up an immunity to the measles virus. To prevent a huge scandal that could severely harm the reputation of the entire discipline of cultural anthropology, a public debate was held at the AAA meeting about the ethical aspects of Chagnon's research practices. According to his critics, he had violated the ethics of ethnographic fieldwork in order to prove his sociobiological hypotheses about the genetic causes of violence and warfare (Turner 2001).

The debate about Chagnon did not only focus on the ethics of field research among a vulnerable group, but also on the professional responsibility of anthropologists. In this context, Chagnon was criticized for collaborating with a group of wealthy Venezuelans in order to obtain access to the living area of the Yanomami Indians, despite the Venezuelan government rejecting his application for a research visa. More importantly, however, Chagnon was criticized for not objecting to the use, or abuse, of his representation of the Yanomami as extremely violent and prone to warfare. Chagnon's characterization of the Yanomami was later used to prevent the establishment of a reservation by gold prospectors who joined into a coalition with politicians, military leaders, and journalists so they could continue their search for gold in the Amazon. A Brazilian organization of anthropologists submitted a form of protest about this to the AAA. This protest, in turn, caused the AAA to investigate the work of Chagnon and its dissemination.

The report of the so-called El Dorado Task Force, however, is equally as controversial as Chagnon's work. Chagnon's critics argue the report is too weak, while his supporters argue it is too strong. The report did rehabilitate the reputation of epidemiologist James Neel, but Chagnon will likely be forever stuck in a widely contested ethical debate. The confusion about the report, however, has only increased since the membership of the AAA rejected it (Borofsky 2005). At the 2009 AAA meeting, a new panel was organized to discuss this controversy, which accused the AAA of scandalous behaviour by using Tierney's book to investigate Chagnon and his companion Neel, rather than defending these researchers against so-called false journalism by Tierney.

Assignment

1. Is it possible to use modern societies as ethnographic analogies to suggest how early prehistoric societies operated? Or should anthropological research always be situated in a specific social, political, and historical context?
2. What guidelines can we suggest to ensure that anthropological field research practices do not violate a code of ethics for research involving human participants?
3. How can we define the professional responsibility of anthropological researchers to influence the reception and use of their findings?
4. Do anthropologists have an obligation to protect the interests of their research participants, even when they are allegedly violent?

Case Study: Fraud or Fiction? Diary of a Teenage Girl

When psychoanalysis was still a developing discipline (early 1900s), a publishing house was founded that would issue psychoanalytic literature exclusively. Among its publications were the journals of a teenage girl named Gretl (*Tagebuch eines halbwüchsigen Mädchens*, 1919) (Fig. 6.6). She was 11 at the start of the journal, and 14 by the end. Her 'case' seemed to illustrate Freud's theories on psychosexual development perfectly.

Gretl came from an upper middle-class family and was a typical teenage girl: she gossiped, quarreled with her friends and made up again, cried hot tears over silly things, and, of course – she came of age. More specifically, she became aware of her own sexuality. She discovered the difference between boys and girls and found out about the 'great secret.' Writing in an October 9th entry she exclaimed: 'Now I know everything!! So that's where little children come from.'

By the time she turned 14, her mother had died. At the funeral, she expressed feelings of hurt because her older sister Dora was allowed to walk besides her father in church, but she was not. Dora even said to her sister that the death of their mother was 'God's way of punishing their father' because they (the sisters) had kept things hidden from their mother – a typical instance of 'magical thinking', as described by Freud.

The diary was supposedly authentic. Not a word was altered, the anonymous editor of the journals assured the reader, nor had grammatical errors been corrected (so presumably slight but meaningful slips of the pen could reveal the young girl's true intentions).

The diary confirmed many psychoanalytic notions in detail (sexual anxiety, childhood jealousy, oedipal feelings, etc.). In fact, in an introductory note to the book, Sigmund Freud wrote: 'The diary is a little gem. I really believe it has never before been possible to obtain such a clear and truthful view on the mental impulses

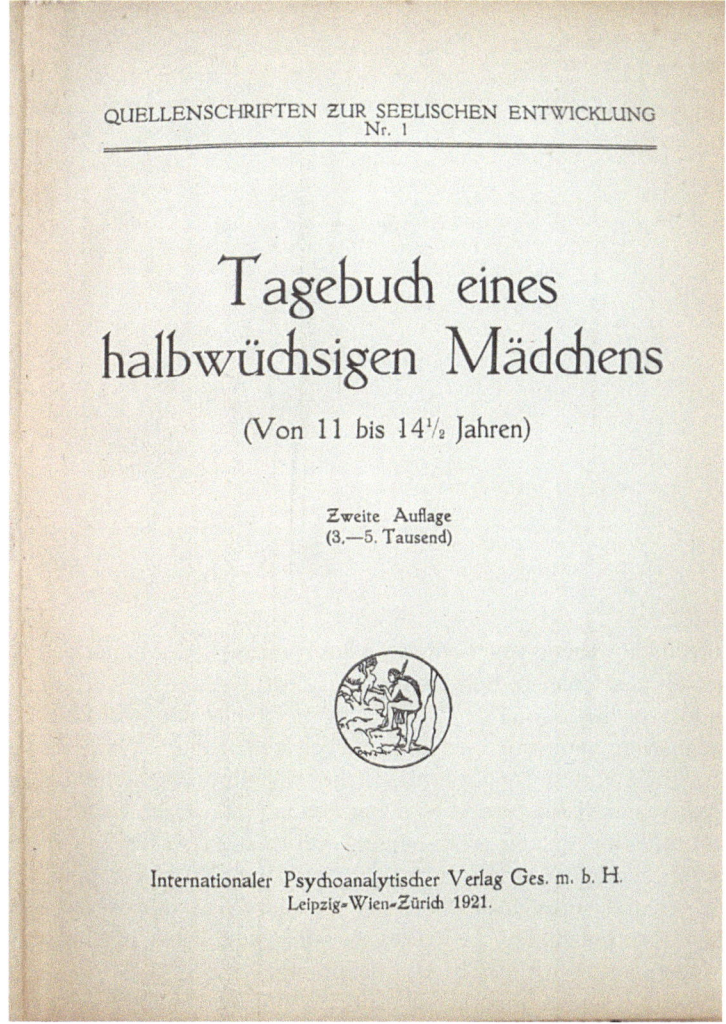

QUELLENSCHRIFTEN ZUR SEELISCHEN ENTWICKLUNG
Nr. 1

Tagebuch eines halbwüchsigen Mädchens

(Von 11 bis 14½ Jahren)

Zweite Auflage
(3.—5. Tausend)

Internationaler Psychoanalytischer Verlag Ges. m. b. H.
Leipzig-Wien-Zürich 1921.

Fig. 6.6 Cover of *Diary of a Young Girl* (Tagebuch eines halbwüchsigen Mädchens), published in 1919

that characterize the development of a girl in our social and cultural stratum the years before puberty.'

A year after the journal's arrival, Hermine Hug-Hellmuth, an early (and now forgotten) follower of Freud and practicing child analyst, confirmed rumors that it was she who had collected the young girls notes and published them (Fig. 6.7). In 1921, an English translation of the diary appeared, and it became a commercial success. Shortly thereafter however, accusations of fraud bubbled to the surface. According to a critic, Gretl's journals were too sophisticated to be true. The critic?

Fig. 6.7 Hermine
Hug-Hellmuth, an early
follower of Freud. (Source:
wikicommons)

Cyril Burt, then a young psychologist, who ironically would later be exposed as a fraud himself (see Chap. 5, Box 5.6 on Cyril Burt).

The editor of the journal denied all allegations, claiming that Gretl's published diary entries were 'authentic' and had not been 'touched up.' While the controversy raged on, Hug-Helmuth tragically died (she was murdered by her nephew, whom she had partly raised and treated with psychoanalysis). In the years after her death, more incriminating details of fraudulent information surfaced. Critics revealed numerous chronological errors, including Gretl's mention of a grading system at her school which was introduced only after the diary had supposedly been written. Today, historians concur that the diaries are not authentic and, in all likelihood, were largely if not entirely made up by Hug-Hellmuth.

This case raises three important questions: (1) Why would someone want to publish a fictitious diary? (2) How did the psychoanalytic community respond to the affair when the diaries were exposed as fraudulent? And (3) How does a case like this reflect on the field of psychoanalysis in general?

It may not have been fame the author was looking for. Rather, as Appignasi and Forrester proposed in their review of the case (1992, p. 200), Hug-Hellmuth had merely meant to 'provide evidence for Freud's theories.' While this explanation is to a certain extent circular, it still gives us a hint as to her possible motives. Psychoanalysis was still a young science in the first few decades of the twentieth century and it was very much in need of confirmation, with many of Freud's followers struggling to find support for his concepts. Lacking a library of psychoanalytic cases in the field's early years, many enthusiasts turned to myths, stories, and historical figures for evidence. The 'Diaries of a Young Girl' seems to fit perfectly into this pattern of early 'missionary work' that was meant to give credibility to psychoanalysis.

How did the psychoanalytic community respond to the allegations? The editors of the publishing house were still making desperate efforts to check the diaries' authenticity by the time Hug-Hellmuth died (in 1924) (Borch-Jackobson and Shamdasani 2012, p. 284). By 1927, the publishers decided to retract the book, directing bookstores to return any remaining copies without an accompanying rationale. The English translation, published in the UK, however, remained available and was reprinted several times, with no note of its fictitious nature. Additionally, a number of practicing psychoanalysts continued to defend the diaries. As an example, Helene Deutsch said she considered Hug-Hellmuth to be 'too imaginative to have recreated a childhood out of whole cloth' (quoted in Roazen 1985, p. 19). In sum, while the history of psychoanalysis is riddled with controversies, it appears that the case of the forged diary had little to no impact on the early reception of the field.

Assignment

Should the discovery of the fraudulent diaries have had a bigger impact on the early reception of psychoanalysis? Consider some of the possible reasons they didn't.

Personal factors – Hug-Hellmuth was a woman working in a field dominated by men; she was not considered a central figure in psychoanalysis.

Contextual factors – Hug-Hellmuth met an untimely death and could never be held accountable, nor could fraud be sufficiently established at the time.

Disciplinary factors – Psychoanalysis has often been accused of being a sect-like cult, not open to discussion.

1. Which of these factors do you think holds the most weight?
2. Think of a similar case of fraud (Diederik Stapel or Cyril Burt for example) and consider which of these factors impacted the field most. How so?

Suggested Reading

For a general introduction into the methodological problems in present-day science, we recommend John Staddon's 2017 highly readable *Scientific Method: How Science Works, Fails to Work or Pretends to Work*. We also recommend Fanelli's papers on retractions in scientific literature, and the question of whether or not they signify a positive trend (Fanelli 2013, 2018). A must read on the subject of false positives can be found in Ioannidis (2005) 'Why Most Published Research Findings Are False.' Finally, we recommend Stroebe and Spears' 2012 article 'Scientific Misconduct and the Myth of Self-Correction in Science,' which provides a crucial discussion of some of the proposed measures to counter the problems discussed in this chapter.

References

Baud, M., Legêne, S., & Pels, P. (2013). *Draaien om de werkelijkheid* [Circling around reality]: Rapport over het antropologisch werk van prof. Em. M.M.G. Bax. Amsterdam University, 9 September 2013.

Chamber, C. (2017). *The seven deadly sins of psychology: A manifesto for reforming the culture of scientific practice*. Princeton: Princeton University Press.

Cromey, D. W. (2010). Avoiding twisted pixels: Ethical guidelines for the appropriate use and manipulation of scientific digital images. *Science and Engineering Ethics, 16*, 639–667. https://doi.org/10.1007/s11948-010-9201-y.

Fanelli, D. (2013). Why growing retractions are (mostly) a good sign. *PLoS Medicine, 10*(12), e1001563. https://doi.org/10.1371/journal.pmed.1001563.

Fanelli, D. (2018). Opinion: Is science really facing a reproducibility crisis, and do we need it to? *Proceedings of the National Academy of Sciences, 115*(11), 2628–2631. https://doi.org/10.1073/pnas.1708272114.

Fiedler, K., & Schwarz, N. (2016). Questionable research practices revisited. *Social Psychological and Personality Studies, 7*(1), 45–52. https://doi.org/10.1177/1948550615612150.

Franco, A., Malhotra, N., & Simonovits, G. (2014). Publication bias in the social sciences: Unlocking the file drawer. *Science, 345*(6203), 1502–1505. https://doi.org/10.1126/science.1255484.

Greenwald, A. G., Pratkanis, A. R., Leippe, M. R., & Baumgardner, M. H. (1986). Under what conditions does theory obstruct research progress? *Psychological Review, 93*(2), 216–229. https://doi.org/10.1037/0033-295X.93.2.216.

Hubbard, R., & Armstrong, J. S. (1994). Replication and extension in marketing: Rarely published but quite contrary. *International Journal of Research in Marketing, 11*(3), 233–248. https://doi.org/10.1016/0167-8116(94)90003-5.

Ioannidis, J. P. A. (2005). Why most published research findings are false. *PLoS Medicine, 2*(8), e124. https://doi.org/10.1371/journal.pmed.0020124.

Ioannidis, J. P. A. (2011). Quantifying selective reporting and the proteus phenomenon for multiple datasets with similar bias. *PLoS One, 6*(3), e18362. https://doi.org/10.1371/journal.pone.0018362.

Ioannidis, J. P. A. (2012). Why science is not necessarily self-correcting. *Perspectives on Psychological Science, 7*(6), 645–654. https://doi.org/10.1177/1745691612464056.

John, L. K., Loewenstein, G., & Prelec, D. (2012). Measuring the prevalence of questionable research practices with incentives for truth telling. *Psychological Science, 23*(5), 524–532. https://doi.org/10.1177/0956797611430953.

Jonas, E., Schulz-Hardt, S., Frey, D., & Thelen, N. (2001). Confirmation bias in sequential information search after preliminary decisions: An expansion of dissonance theoretical research on selective exposure to information. *Journal of Personality and Social Psychology, 80*(4), 557–571.

Köbben, A.J.F. (2012). *Bedrog in wetenschap* [Fraud in science] Lecture before the department of humanities at the Royal Academy of Sciences, 9 January 2012.

Kolfschooten, F. (2012). *Ontspoorde wetenschap. Over fraude, plagiaat en academische mores over fraude, plagiaat en academische mores [science derailed. On fraud, plagiarism, and academic Morales]*. Amsterdam: Uitgeverij de Kring.

Loftus, E. A., & Ketcham, K. (1994). *The myth of repressed memory*. New York: St Martin's Press.

Macpherson, R., & Stanovich, K. E. (2007). Cognitive ability, thinking dispositions, and instructional set as predictors of critical thinking. *Learning and Individual Differences, 17*(2), 115–127. https://doi.org/10.1016/j.lindif.2007.05.003.

Masicampo, E. J., & Lalande, D. R. (2012). A peculiar prevalence of p values just below .05. *The Quarterly Journal of Experimental Psychology, 65*(11), 2271–2279. https://doi.org/10.1080/17470218.2012.711335.

Møller, A. P., & Jennions, M. D. (2001). Testing and adjusting for publication bias. *Trends in Ecology & Evolution, 16*(10), 580–586. https://doi.org/10.1016/S0169-5347(01)02235-2.

Munafò, M., Nosek, B., Bishop, D., et al. (2017). A manifesto for reproducible science. *Nature Human Behaviour, 1*(1), 0021. https://doi.org/10.1038/s41562-016-0021.

Osborne, J. W., & Overbay, A. (2004). The power of outliers (and why researchers should ALWAYS check for them) check for them. *Practical Assessment, Research, and Evaluation, 9*(6), 1–8.

Perkins, D. N. (1985). Postprimary education has little impact on informal reasoning. *Journal of Educational Psychology, 77*, 562–571.

Pezdek, K., & Banks, W. P. (1996). *The recovered memory/false memory debate*. San Diego: Academic.

Pfeiffer, T., Bertram, L. & Ioannidis, J.P.A. (2011, March 29). Quantifying Selective Reporting and the Proteus Phenomenon for Multiple Datasets with Similar Bias. *PlosOne*, https://doi.org/10.1371/journal.pone.0018362

Resnick, B. & Bellus, J. (2018, October 24). *A top Cornell food researcher has had 15 studies retracted. That's a lot*. Retrieved from: https://www.vox.com/science-and-health/2018/9/19/17879102/brian-wansink-cornell-food-brand-lab-retractions-jama

Sharaf, M. (1983). *Fury on earth: A biography of Wilhelm Reich*. New York: Da Capo Press.

Shemer, M. (2002). *The borderlands of science. Where sense meets nonsense*. Oxford: Oxford University Press.

Simmons, J. P., Nelson, L. D., & Simonsohn, U. (2011). False-positive psychology: Undisclosed flexibility in data collection and allows presenting anything as significant. *Psychological Science, 22*(11), 1359–1366. https://doi.org/10.1177/0956797611417632.

Staddon, J. (2017). *Scientific method: How science works, fails to work or pretends to work*. London: Taylor & Francis.

Stroebe, W., Postmes, T., & Spears, R. (2012). Scientific misconduct and the myth of self-correction in science. *Perspectives on Psychological Science, 7*(6), 670–688. https://doi.org/10.1177/1745691612460687.

Teixeira da Silva, J. A. (2015). Debunking post-publication peer review. *International Journal of Education and Information Technology, 1*(2), 34–37. http://www.aiscience.org/journal/ijeit.

Toplak, M.E. & Stanovich, K.E. (2003), Associations between myside bias on an informal reasoning task and amount of post-secondary education. Appl. Cognit. Psychol., 17: 851–860. doi: "https://doi.org/10.1002/acp.915" https://doi.org/10.1002/acp.915.

Van Assen, M. A. L. M., van Aert, R. C. M., Nuijten, M., & Wichert, J. M. (2014, January 17). *Why publishing everything is more effective than selective publishing of statistically significant results*. PlosOne. https://doi.org/10.1371/journal.pone.0084896.

Wolfe, C. R., & Britt, M. A. (2008). Locus of the myside bias in written argumentation. *Thinking & Reasoning, 14*, 1–27.

References for Case Study: Yanomami Violence and the Ethics of Anthropological Practices

Borofsky, R. (2005). *Yanomami: The fierce controversy and what we can learn from it*. Berkeley: University of California Press.

Chagnon, N. A. (1983). *Yanomamö: The fierce people* (3rd ed.). New York: Holt, Rinehart, & Winston. (Original work published 1968).

Ferguson, R. B. (1995). *Yanomami warfare: A political history*. Santa Fe: School of American Research Press.

Tierney, P. (2000a). The fierce anthropologists. *The New Yorker*, October 9, pp. 50–61.

Tierney, P. (2000b). *Darkness in el dorado: How scientists and journalists devastated the Amazon*. New York/London: Norton & Company.

Turner, T. (2001). *The Yanomami and the ethics of anthropological practice*. Ithaca, NY: Cornell University, sponsored by the Latin American Studies Program, with support from the U.S. Department of Education Title VI.

References for Case Study: Fraud or Fiction? Diary of a Teenage Girl

Appignasi, L., & Forrester, J. (1992). *Freud's women*. London: Weidenfeld & Nicolson.
Borch-Jackobson, M., & Shamdasani, S. (2012). *The Freud files*. Cambridge: Cambridge University Press.
Hug-Hellmuth, H. (1919). *Tagebuch eines halbwüchsigen Mädchens*. Vienna: Internationaler Psychoanalytischer Verlag.
Roazen, P. (1985). *Helene Deutsch. A psychoanalyst's life*. New York: Meridian Books.

Part III
Ethics and Trust

Chapter 7
Confidentiality

Contents

Electronic Supplementary Material: The online version of this chapter (https://doi.org/ 10.1007/978-3-030-48415-6_7) contains supplementary material, which is available to authorized users.

© The Author(s) 2020
J. Bos, *Research Ethics for Students in the Social Sciences*,
https://doi.org/10.1007/978-3-030-48415-6_7

After Reading This Chapter, You Will:

- Know what confidentiality entails and why it is important in research
- Become familiar with techniques of securing confidentiality
- Understand the difference between anonymity and confidentiality
- Be able to identify breaches of confidentiality

Keywords Anonymity · Autonomy · Blind protocols · Confidentiality · Data leakage · Deductive disclosure · Dignity · Hacking · Identifiers · Informed consent · k-anonymity · Personal data · Privacy (attacks) · Proxy consent · Re-identification · Research data management plan [RDMP] · Self-disclosure · Trust

7.1 Introduction

7.1.1 The Privacy of Facebook

Since the launch of Facebook as a (commercial) social media platform, its potential as a treasure trove of data on the dynamics of social networks and both online and offline behavior was quickly recognized by sociologists. In 2006, just a few years after Facebook entered the public sphere, a group of researchers downloaded the publicly available data for an entire cohort of some 1700 freshmen students at an undisclosed US college. The data haul included demographic, relational, and cultural information for each individual, and the interested sociologists intended to use it in generating multiple future research projects.

The researchers had requested and obtained permission to utilize the data for research purposes from Facebook, the college involved, as well as the college's Institutional Review Board (IRB). Notably, they did not seek consent from the individual users, although steps were taken to ensure that the identity and privacy of the students remained protected, including the removal of identifying information (such as the names of the students). Also, they demanded that other researchers who wished to use the data for secondary analysis would sign a 'terms and conditions for use' agreement that prohibited any attempts to re-identify the subjects. Convinced that this would ensure confidentiality, the data set was released in 2008 as the first installment of a data sharing project purported to run until 2011.

However, just four days after the data's release, Fred Stutzman, a PhD student, questioned the non-identifiability of the data, writing: 'A friend network can be thought of as a fingerprint; it is likely that no two networks will be exactly similar, meaning individuals may be able to be identified in the dataset post-hoc' (quoted in Zimmer 2010, p. 316). Soon thereafter, it was established that the source of the data was very likely Harvard College, and although no individual subjects were identified at that point, the dataset was taken offline as a precaution.

In a discussion of this case, Zimmer (2010) observed that the researchers who initialized the project made two assumptions. Firstly, they believed that even if the data set were 'cracked' (allowing individual subjects to be identified), the privacy of the subjects would not be violated because the data was already public in the first place. Secondly, they assumed the research ethics implications had been sufficiently observed by consulting the college's IRB and taking steps to anonymize the data.

Addressing both arguments, Zimmer argued that extensive collection of personal data over a long period of time, made publicly available for the purpose of social networking only, by subjects whose consent was neither sought nor received *does* constitute a violation of their privacy (Box 7.1). Additionally, Zimmer found it to be a breach of research ethics because subjects were not provided access to view the data to correct for errors or request the removal of unwanted information (for further discussion of this case, see Zimmer 2010) (Fig. 7.1).

This case raises two important issues. The first being that confidentiality is not merely a matter of shielding research participants' identities. Confidentiality is about knowing what sort of personal data may be available, to whom, and under which conditions – in essence, it's about considering the whole picture. It also implies the participant's right to being informed about the scope of their participation, to being explicitly asked to take part in the research project, and extends to their right to retain (some degree of) control over their own data. Secondly, this case exemplifies how quickly, and recently, our understanding of confidentiality has changed. Not only is it very unlikely that an IRB would approve of the above procedures today, but Facebook and other online social networks have also been increasingly critiqued for their defective privacy policies, of which we have only recently become aware.

In this chapter, we outline confidentiality from a wide lens, and flesh out some of its most salient properties. We will discuss some difficulties with securing confidentiality and examine its relationship to anonymity and informed consent procedures. Finally, we discuss breaches of confidentiality and their consequences.

Box 7.1: What Is Personal Data?
What is defined as 'personal' may differ from one person to the next, although there are some obvious instances that perhaps everyone would agree is personal, such as your medical history, sexual orientation, or certain beliefs or opinions. Research policies distinguish between these various categories of 'personal data.' The following list, derived in part from the European General Data Protection Regulation, is not exhaustive (Fig. 7.2).

Full name	Sexual orientation
Home address/email address/IP address	Trade union membership
Date and place of birth	Passport/ID/driver's license number
Mother's maiden name	Credit card number
Ethnicity/race	Telephone number
Age	Job position
Religious or philosophical beliefs	Biometric records
Political opinions	Criminal record

Fig. 7.1   Can you tell me something personal?

Fig. 7.2   *What are personal data*? (Source: European Commission, data protection)

We restrict our analysis, as we have in all other chapters in this book, to research ethics, and do not cover confidentiality issues within professional relationships, and only briefly touch on the (often severe) judicial components of the issue.

7.2   Defining Confidentiality

7.2.1   What Is Confidentiality?

Any information relating to the private sphere of a person that they wish not be shared with others is considered 'confidential.' This information is differentiated from 'public information,' which everyone has a right to access. The right of research participants to not disclose certain information and to retain control over their privacy has increasingly been acknowledged inside and outside of academia and has become subject to extensive legislation.

In research ethics, the crucial principle of confidentiality entails an obligation on the part of the researcher to ensure that any use of information obtained from or shared by human subjects respects the *dignity* and *autonomy* of the participant, and does not violate the interests of individuals or communities (see Box 7.2 for clarification of concepts). The right to confidentiality in research is recognized in international bio-ethical guidelines, such as the 'Helsinki Declaration' (last updated in 2013), and the European General Data Protection Regulation (GDPR, effective 2018).

In practice, safeguarding confidentiality entails that the researcher observes the following restrictions:

- Research participants remain anonymous by default
- Researchers do not obtain private data unless there is good reason to
- Participants must be briefed on the goal or purpose of the research, its means of investigation, and who has access to the data
- Participants must give active consent, are not coerced to participate, and retain the right to withdraw their cooperation at any moment (even after the study has been completed)
- Participants must be provided with an opportunity to review their data and correct any mistakes they perceive

---

Box 7.2: 'Privacy, Autonomy, Confidentiality, Dignity'
Autonomy: the capacity to make uncoerced decisions for oneself.
Privacy: an individual's sphere of personhood, not open to public inspection.
Confidentiality: private information that a person may not want to disclose.
Dignity: a sense of one's personal pride or self-respect.

7.2.2 Confidentiality and Trust

Confidentiality pertains to the understanding between the researcher and participant that guarantees sensitive or private information will be handled with the utmost care. Ultimately, confidentiality is rooted in trust.

The participant must trust that the researchers will fulfill their responsibilities and protect the participant's interests. To ensure this happens, an agreement is drawn up in which these duties are specified and communicated to the participant (see Sect. 7.3).

In online and computer-assisted research – a variety that often lacks a face-to-face dimension and perhaps poses a greater privacy threat than traditional research – trust and especially the perception of control over one's own information are key. Addressing the concerns dictates how much and to what degree participants are willing to disclose about themselves (Taddei and Contena 2013).

7.3 Securing Confidentiality

7.3.1 Informed Consent

Perhaps the most important instrument for securing confidentiality is the *informed consent procedure*. It is rooted in the idea that involvement in research should have no detrimental effects on the participants, honor the individual's fundamental rights, and respect relationships, bonds, and promises.

Certain conditions and arrangements have been designed to guarantee safe participation in research. These procedures assume the shape of a contract with a participant who actively and knowingly agrees with the conditions. *Informed consent* typically notifies the participant of the following items:

- Name(s) and affiliation of researcher(s)
- Goal or aim of the research (in comprehensible language)
- Research techniques or procedures to which the participant is subjected
- Risks involved (if any)
- Estimate of time investment
- Agreement on compensation (if any)
- Conditions of confidentiality (anonymization or pseudonymization)
- Storage, usage, and access to data
- Rights of the participant:
    - to withdraw at any moment
    - to review/correct erroneous data (if possible)
    - to receive/be informed about the results (if interested)

- Complaint procedures (including contact details of an independent commission or officer)

Informed consent procedures have become mandatory in social scientific research for qualified researchers, including PhD candidates. Undergraduate students, who do research under the supervision of qualified staff, are generally also required to make use of these procedures (with the responsibility for their proper implementation that of the supervisor). Many of these principles are paralleled by similar procedures in medicine, law, and other professional domains (for further discussion, see Bok 1983, and Israel 2014).

7.3.2 Difficulties with Informed Consent

While informed consent thus aims to protect the participant, a few difficulties arise with how we approach it, some of a philosophical nature, others more practical.

One contention is that informed consent is biased towards a particular (Western) view of individuality. Research participants are supposed to be autonomous, well informed, capable subjects who are solely responsible for their own behavior and for knowing with whom formal contracts can be negotiated, and understanding the conditions of their participation.

Not all participants fit into this ideal of autonomous agency. Children (minors), vulnerable communities (for example those who harbor suicidal ideation), or anyone in a dependent relationship who may not be (entirely) free to refuse participation, as well as those who may be unable to fully understand the 'contract,' all fall outside of this ideal of autonomous agency.

Furthermore, participants may not always be in the position to appreciate exactly what the conditions of participation entail. This is exacerbated when deception is involved, or when the research design does not allow for the participant to be fully or correctly informed about their role in the study.

Finally, confidentiality procedures warranting subject autonomy favor quantitative research (experimental studies, surveys) that does not require meaningful relationships to be formed with participants. In qualitative research (interviewing, participant observations, etc.) these relationships are pivotal, and formal agreements, such as informed consent procedures, 'can be problematic in a culture that values relationships over roles and position' (LaFrance and Bull 2009, p. 145).

Although it is possible to address some of these difficulties in the informed consent agreement between researcher and participant, other issues remain unresolved, especially those regarding qualitative research, to which we return below.

In the final chapter of this book, we review the procedural dimension of confidentiality. There we discuss how to weigh the various risks and benefits, explore how to deal with deception, discuss how to assess the vulnerability of participants and intrusiveness of research, and what to do with 'chance findings.'

7.4 Anonymity

7.4.1 Anonymity Versus Confidentiality

These two concepts, *anonymity* and *confidentiality*, are related, but differ in some important respects. Anonymity can be defined as the degree to which the source of a message can be identified (Scott 1995). It ranges from very high (source is nearly impossible to identify) to none (source is easily identifiable or in fact already identified). Confidentiality, on the other hand, relates to an agreement between the researcher and the participant. The former concerns the initial collection of data, the latter makes promises to not disclose specific personal information.

Seeing as how researchers need to protect the participant from unwanted consequences, anonymity seems a safer guarantee for achieving this goal than confidentiality. A researcher who offers *anonymity* does not record any identifying information. If *confidentiality* is offered, however, identifying information is recorded, but this information will not be disclosed to others.

Does it matter much whether you offer anonymity or confidentiality to your participants? Whelan (2007) demonstrated that research participants are aware of the difference and are equally able to appreciate the different degrees of protection offered under both conditions. This raises the question of whether 'weaker' confidentiality agreements could undermine the reliability of research. In a comparative analysis (comparing an anonymous and a confidential condition) of self-reported substance abuse surveys among 15 and 16-year-old Icelandic students, Bjarnason and Adalbjarnardottir (2000) found no evidence that a confidential condition lowered the study's reliability. Conversely, Lelkes et al. (2012) found that complete anonymity may compromise self-reporting.

Anonymity thus may be a more absolute, though not 'better,' criterion than confidentiality in ensuring the participant's right to privacy. Confidentiality, on the other hand, allows for the creation of a relational dimension that is explicitly left out in anonymity. The importance of relationships in research is a ripe field of study (Box 7.3).

In brief, there can be good reason to offer confidentiality as opposed to anonymity, although anonymity is generally preferred.

7.4.2 Managing Anonymity

While anonymity is the norm in research, safeguarding it poses an increasingly prevalent challenge in our digital age. 'Privacy attacks' and 'data leakages' are rampant and the mechanisms for using public data for participant re-identification have greatly increased (Ramachandran et al. 2012). Netflix's 2019 true crime documentary 'Don't F*ck with Cats' gives an instructive illustration of how it is possible to identify an anonymous individual from a YouTube video by combining contextual

Box 7.3: Breaking Confidentiality in Good Faith? A Dilemma
Consider the case of a student who did research into 'workplace inclusion' at a large governmental institution. The student was commissioned to research the attitudes and experiences of employees with workplace inclusion.

Using a qualitative design, the student interviewed some 20 participants in three different departments in the institution. In accordance with standing institutional policy, the student was not allowed to identify participants on basis of their ethnicity (employees were not ethnicity registered at the institution). However, during the student's research, she found that ethnicity *did* play a role in how employees experienced feelings of inclusion and exclusion. Some participants felt that 'the fact that they belonged to an actual or perceived group determined how they were treated by fellow employees and the managers at the institution.'

This result was clearly of importance for the study, yet it conflicted with institutional policy that did not allow the student to identify the ethnic background of the participants. A dilemma arose on how to continue. Should she, or should she not mention ethnicity?

How would you advise the student to proceed? Should the student make use of this information and break confidentiality of the basis that she acts in good faith, or should all mention of ethnicity be removed, in accordance with institutional policiy, at the cost of losing vital information?

(Case was communicated to the author. Quotes are altered to prevent identification.)

information in the video (type of electoral receptacles, doorhandles, background noises), publicly available records (street maps, location of shops, etc), and the use of common sense.

Such easy, cheap, and powerful re-identifications not only undermine our faith in anonymization and cause significant harm, they are also difficult to avoid (Ohn 2010). The advantages of digitalization, including increased potential to collect, process, analyze, store, and share data, are countered by new privacy risks, in particular the disclosure of personal data and re-identification. And although GDPR is meant to avoid these risks as much as possible, Rhoens (2019, p. 75) warns how in the age of 'big data', power tends to be shifted towards data controllers, reducing consumers' autonomy, undermining a key element of private law in Europe.

In health-related research there is the ever-present risk that databases get hacked, which are full of sensitive information regarding symptoms, diagnoses, and treatment plans. Longitudinal studies (which follow (groups of) individuals over a long period of time) must allow for an identifying key at least until the study is finished, and thus pose the risk that while the study runs the key is revealed. Online social network analyses that deal with large amounts of data run the risk that the privacy of its users may be infringed upon (as the Facebook example demonstrated). Lastly, as studied by Williams and Pigeot (2017), we should be wary of powerful organizations, corporations, and governments, who are gathering vats of information about us, further arguing that 'We have good reasons to fear that this may damage our interests, since their processes of data gathering are so shadowy and unaccountable' (p. 248).

In an attempt to prepare for privacy attacks and possible data leaks today, research institutions require that anonymization be part of a larger *research data management plan* that invokes policies about data security, data privacy, and data licensing (see Patel 2016) (Box 7.4).

7.4.3   How to Secure Anonymity?

Though this question regards research techniques rather than research ethics, we will have to outline the constraints of this issue before we can discuss the ethical aspects related to it (Fig. 7.3).

The anonymization of data necessitates that *identifiers* (defined below) are changed, either by removing them outright, or by substitution, distortion, generalization, aggregation, or the employment of any number of masking techniques (Ohm 2010).

Direct identifiers, such as name, address, or zip code, can be removed or substituted by a pseudonym or code without loss of information. If substitution by a code is used, a key that allows reidentification may be kept, but as explained above, that key can subsequently become a security risk.

Indirect identifiers, such as information regarding occupation, training, age, or workplace can be aggregated or generalized, but therein lies the risk of information loss (for example, when the value '19' is substituted for the aggregated value 'between 15-20 years old').

Box 7.4: Research Data Management Plan (RDMP)
Any RDMP must be compliant with applicable national or international standards and stipulate conditions for the following data-related considerations (pertaining to both new data and amendments of existing projects):

- security, privacy protection, and transparency
- working with sensitive data
- archiving of research data
- storage procedures
- creating metadata
- authorship of data and data use
- verifiability of data
- searchability of data
- data sharing and licensing
- retention period and contact details of the data manager

(Compiled after various university library sources)

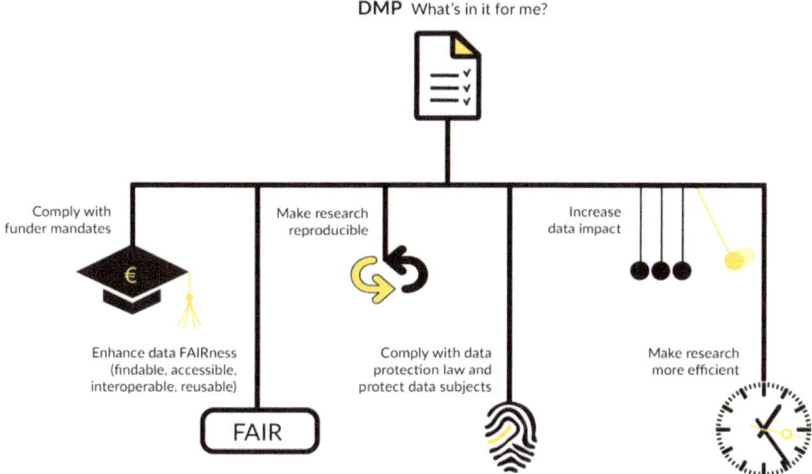

Fig. 7.3 *Research data management plan*. (Source: Utrecht University, research data management support)

Quasi-identifiers make it possible that a participant be identified when disparate information that by itself do not identify a subject are combined to create a clearer picture. For example, in an institution, the combination of age, position, and gender may lead to the identification of the participant if there is only one person with that specific set of characteristics.

In order to anonymize sets of data while retaining as much original information as possible, certain techniques have been developed. One known as *k-anonymity* was specifically designed for quantitative data sets (introduced by Samarati and Sweeney 1998, and since improved, see Amiri et al. 2016).

This technique allows for sensitive data to be recorded but disallows that data may be combined to create quasi-identifiers. Essentially, k-anonymity requires that there always be several individuals that match *any* combination of values in the same set of data (see Doming-Ferrer and Torra 2005; Ciriani et al. 2008, for further discussion of k-anonymity and Zhou et al. 2008, for a comparison with other anonymization techniques) (Fig. 7.4).

7.4.4 Is Complete Anonymization Possible?

The answer to this question is… probably not. The main reason being that anonymizing techniques, including k-anonymity, do not offer fool proof protection against the malicious use of background information, data triangulation, or even just basic web searches (Martin et al. 2007). As a result, 'deductive disclosure' (Tolich 2004) occurs, where certain individual or group traits make them identifiable in research reports.

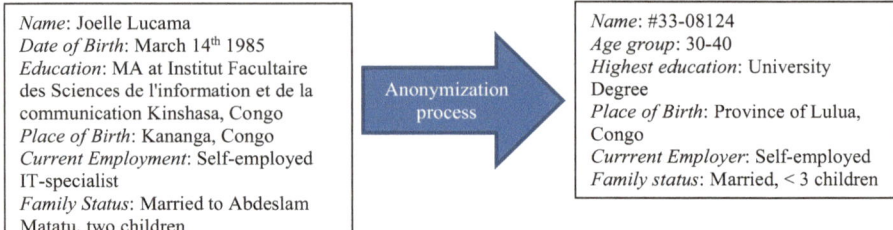

Fig. 7.4 Anonymization: use of coding, aggregation and redaction (fictitious case). Adapted from Ruth Gerathy, *Anonymisation and Social Research* (2016)

For example, user profiles for many common web-oriented services (such as Facebook or Google profiles) provide a host of background information that make it easy to re-identify research participants in anonymized data sets (Kumpošt and Matyáš 2009). Also, with the aid of publicly available census data that contains records of individual's birth date, gender, and address, *quasi-identifiers* can be constructed, and anonymized records from smart meter data (Buchmann et al. 2013) or cell phone users (Zang and Bolot 2014) can be used together to re-identify anonymous research participants. Similarly, anonymized online social networks have been de-anonymized with the aid of certain re-identification algorithms (Narayanan and Shmatikow 2009).

The reality is that at this moment, anonymization is not absolute, but a matter of degree (Dawson 2014). A dataset may never be completely safe from intentional attacks, and therefore re-identification of anonymized data presents serious policy and privacy implications (Lubarsky 2017; El Emam et al. 2019).

7.4.5   Anonymization in Qualitative Research

Qualitative research is performed within a diversity of disciplines such as ethnography, anthropological field work, community studies, as well as clinical interviews and (critical) discourse analysis (the study of larger connected sets of textual corpuses). A defining feature of this form of research is that it deals with texts and their non-quantifiable characteristics; the heterogenous and ambiguous structure of language.

When compared to quantitative research, qualitative researchers are confronted with both similar and different challenges regarding anonymization, which we will now explore.

What is *similar* is that qualitative researchers also must consider confidentiality. What personal information are they allowed to make public (with consent from the participant), and what is off limits? When qualitative researchers choose to remove or alter identifiers, they too must worry that background knowledge will allow online sleuths to re-identify (some of) the participants. But masking the identity of an interviewee or a patient may be even more difficult because of the wealth of self-disclosing information available online.

An additional comparable difficulty that quantitative researchers must also resolve when anonymizing their data, is that even when direct and indirect identifiers are removed, *contextual identifiers* in an individuals' narrative remain. For example, certain (unusual) life events, particular details or circumstances, and specific places are all difficult to anonymize without distorting the text. Despite this difficulty, attempts have been made to create computer programs that facilitate the finding and replacement of personal identifiers (see Kaiser 2009 for a discussion).

What is *different* in qualitative research is that not all researchers share the 'fetish of individualism' (Weinberg 2002). Some insist that research communities are places where people know each other, where ideas and knowledge is shared, and where no-one goes unnoticed. For this reason, they argue, anonymity is virtually unachievable, and the question is whether anonymity is even desirable at all (van den Hoonaard 2003; Walford 2005; Tilley and Woodthorpe 2011). Other researchers have argued to the contrary, and insist that in spite of these objections, anonymity should still prevail as a guarantee against gratuitous and casual identification which does nothing to add to public understanding (Kelly 2009).

Another notable difference with quantitative research is that the 'situatedness' of qualitative data (Thomson et al. 2005) makes secondary use questionable (use of the same data by different researchers). Qualitative data is 'generated through personal interactions with participants, involving a high degree of trust and a duty of care towards participants' data on the part of the researcher' (Irwin 2013, p. 297). Trust and duty cannot be transferred onto unknown researchers just like that.

Finally, Giordano et al. (2007) point out that sometimes participants specifically wish to be heard and do not want to remain anonymous. Depriving them of a voice that provides personal meaning would deny them of (at least a part of) their autonomy. Giordano proposes that participants be offered a choice of disclosing their identity or not.

In light of the discussion above, consider the following study by Wiles et al. (2006). They conducted research about the use of consent procedures among social scientists and health researchers working with vulnerable populations. The participants – experienced researchers who themselves used qualitative methods – were mostly critical of informed consent procedures. Some had little or no experience with consent forms and were put off by the idea of using them. Others refused point blank to sign the forms Wilkes and her colleagues gave them, which they disqualified as an 'overly formalistic and paternalistic enforcement of a biomedical model' (p. 286).

A difficulty was that some of participants were well known in their field and could easily be identified by others in the research community. They refused to give consent that their data be archived. They also insisted that for reason of indefinability, entire sections of the transcripts be deleted. This meant the loss of important findings, while also making secondary analysis impossible.

The 'researching the researchers' study by Wiles et al. (2006) led to the conclusion that in qualitative research, two items of crucial importance cannot be managed by consent procedures: *trust* of the participant in the research project and their *motivation* to participate in it.

7.5 Breaches of Confidentiality

7.5.1 What Constitutes a Breach of Confidentiality?

A breach of confidentiality means that the obligation of the researcher to ensure their research respects the *dignity* and *autonomy* of the participant is not fulfilled or honored, or that an essential element in the agreement between researcher and participant is broken.

For example, when a promise of anonymity is revoked, then not only is the participant's trust violated, but in the case of any negative consequences the participant may face, the researcher may be liable.

However, *not all breaches are reprehensible*. Some may even be considered justifiable, for example when a higher goal is served. Other breaches may be brought about by a third party and are not a result of the researcher's actions. Or there is the possibility that the breach could simply result from the wish of the participant not to remain anonymous (waiver of confidentiality) (Fig. 7.5).

In the coming section, we discuss examples of these four classifications of breaches in further detail, and identify a number of consequences and possible remedies.

7.5.2 Culpable Breach of Confidentiality

When sensitive, personal, or identifying information from participants is made public without their consent, and it has negative consequences for the participant (or the community), the researcher can be held responsible if they could have prevented this from happening.

Suppose a researcher interviews a number of employees from several different companies about their job satisfaction. The participants are guaranteed complete anonymity. At some point in time a report on the findings is published. Now consider that a supervisor at one of the participating companies reads the report and is able to ascertain a certain participant as one of their employees, based on a number of characteristics. Since this employee has been found to be critical of the

Fig. 7.5 Classification of breaches of confidentiality

Culpable breach of confidentiality	Justifiable breach of confidentiality
Waiver of confidentiality	Enforced breach of confidentiality

organization, the supervisor decides not to promote them. Now, the question can be asked: is the researcher responsible or even liable for the damage?

The answer depends on whether the researcher has *done enough* to secure the anonymity they guaranteed. For example, if only the participants' names were anonymized, but certain unique characteristics remained (gender of the participant, the position in the organization, age group), allowing for easy re-identification, then the researcher might indeed be liable. But if the researcher has done everything that reasonably could be expected from them, and the supervisor deduced the identity of the employee by chance, the breach of confidentiality could be considered merely lamentable, not culpable.

In 2015, the journal *Expert Systems and Applications* published a paper that used several sentences taken from the logged-in section of a website called 'PatientsLikeMe'. One particular quote from an HIV-positive user on the site contained specific dates and types of infections the patient had experienced. This led to a complaint to the editors of the journal that 'a search within PatientsLikeMe for this string [of information], or fragments of it, would quickly identify this patient.' The editors of *Expert Systems and Applications* accepted the validity of this complaint and withdrew the paper. The authors were requested to delete the incriminating quotations and when completed, the paper was later republished (case taken from 'Retraction Watch,' September 2016).

7.5.3   *Justifiable Breach of Confidentiality*

There can be good reason to disclose sensitive or personal information without consent if circumstances force the researcher's hand. This can be found, for example, if a third party could find themselves in immediate or future risk were certain information to remain unknown to them (Box 7.5).

The well-known 1974 'Tarasoff Case' may be taken as the classic example of a justifiable breach of confidentiality, even though it is more of an instance of professional ethics rather than research ethics (see Bersoff, 2014 for a discussion). In the 'Tarasoff Case,' a patient confided in a mental health professional his intentions to kill someone. The intended victim was not alerted and indeed, was later murdered by this patient. When the case came before a court of law, it was ruled that client-therapist confidentiality *should* have been breached because of a higher duty, the protection of the intended victim (see Herbert 2002 for a discussion of the case).

In social science research, analogous situations may present themselves, even though they are rarely as extreme as the Tarasoff Case (see Duncan et al. 2015). For example, Jonathan Kotch (2000) discussed some of the challenges researchers encountered in studying the complex matter of longitudinal child maltreatment research, which led to serious deliberations regarding breaching confidentiality.

The researchers were interested in the behavior of mothers, but in the process they not only collected confidential information about children, but also from them. Would this make these children automatically research participants? And if

Box 7.5: Breaching Confidentiality or Not? A Dilemma

George is a psychologist who is interested in high-risk sexual behavior among adolescents. He speaks with 25 participants, one of whom is Martin, an 18-year-old male who has agreed to be interviewed on the provision of complete confidentiality. During the interview, Martin reveals that he has been diagnosed with HIV and has not informed his partner, even though they have regular unprotected sexual intercourse.

George is worried that he is obliged to breach confidentiality and disclose this information to Martin's partner. For guidance, he consults the *Ethical Principles of the Psychological Association.* It states that confidential information can be disclosed (without the consent of the individual) 'when mandated or permitted by law for a valid purpose such as to protect the client, patient, psychologist, or others from harm' (quoted in Behnke 2014).

The laws in George's country aren't very clear about this issue, though. HIV is a contagious disease but doesn't pose an imminent risk of death, though being infected could be deemed considerable harm.

Are there sufficient grounds for George to breach confidentiality? Argue from one of the following positions:

1. George should inform Martin's partner and does not have to inform Martin about this breach of confidentiality because the partner may be in immediate danger.
2. George should inform Martin's partner but also inform Martin about this breach of confidentiality.
3. George should urge Martin to inform his partner but does not have to interfere himself.
4. George should not interfere in any way as he is bound by confidentiality and the responsibility is Martin's alone.

(Case adapted after Hook and Cleveland 1999)

so, under which conditions could they be considered 'participant' in the research? Logically, parents would have to give consent on behalf of their children (this is called 'proxy consent'), on the presumption that they act in the best interest of their children. But that may not be likely in the case here, given that the research was on child abuse and neglect.

Confidentiality issues were further complicated when suspicion of child abuse arose. Under US law, anyone who suspects maltreatment of a child is legally required to report it. Under these circumstances, is longitudinal research on child maltreatment possible at all?

If the answer is yes, then whose interests prevail: those of the mother, the child, or the researcher? Kotch (2000) argues that all three must be factored in when moving forward with a research project, but they carry different weights. Kotch contents that a child's participation on a basis of 'proxy consent' is ethical as long as the benefits (child welfare, possible beneficial research outcomes) outweigh the risks (harm done to the child). If the child's welfare is at stake, confidentiality may justifiably be breached, but this must be considered very carefully, and weighed against the consequences. This is because the consequences can be substantial, both for the mother (social, personal, and even job-related repercussions) as well as the child (embarrassment, emotional distress).

In order to sensibly handle confidentiality, a special 'blind protocol' was designed for this case, that allowed the mother to respond in writing to sensitive questions that might lead to a suspicion of abuse or neglect, without the interviewer being aware of the answer. Only the principal researchers (PI) would be allowed to review this sensitive material and only they could decide (after careful deliberation) that a case needed to be reported (they eventually did so in five cases out of 442, one of which was confirmed).

7.5.4   Enforced Breach of Confidentiality

There are only a few circumstances that could force a scientist to breach confidentiality. One of those is the enforcement of state regulations. Rik Scarce was a PhD student at Washington State University doing research on an environmental movement in the United States. In his research, he conducted interviews with environmental activists in this movement. Even before his dissertation was published, one of his interviewees attracted the interest of the police. They requested that Scarce appear at the campus police station, where he was interviewed. When he refused to answer certain questions about his research participants and did not allow the police access to his data on grounds of confidentiality, he was subpoenaed to appear before a grand jury. In his testimony, he declared the following: 'Your question calls for information that I have only by virtue of a confidential disclosure given to me in the course of my research activities. I cannot answer the question without actually breaching a confidential communication. Consequently, I decline to answer the question under my ethical obligations as a member of the American Sociological Association [...]' (Scarce 1995, p. 95). This defense was not accepted. Confidentiality simply did not matter to the court, Scarce later wrote (1995, p. 102). He was found in contempt of court and held in custody for over five months.

Although no general conclusions regarding what to do is such cases may be drawn from this case, because laws with respect to liability differ in every country, students should be advised to ensure that their research proposals are in accordance with university policy. In case of doubt, they may want to consult their IRB. See Box 7.6 for further considerations.

Box 7.6: The Russel Ogden Case
In 1994, Russel Ogden, a Canadian MA student in criminology at Simon
Fraser University (SFU), completed a controversial study on assisted suicide
and euthanasia of persons with AIDS, which was illegal at the time, and
attracted a high amount of media attention.

Shortly after the defense of his thesis, based on interviews with
people involved in this activity, Ogden was subpoenaed by the Vancouver
Regional Coroner to reveal his sources. Ogden refused on grounds that he had
promised confidentiality, and that he had acted in accordance with universities policy.

It is noteworthy that Ogden had actively sought approval from the university's independent IRB, noting that anonymity and confidentiality would be
assured with each participant. He also informed his participants in a consent
letter that the 'proposed research project involves data about illegal behavior,'
and that participants would not be required to give information regarding their
identity. Finally, Ogden sought advice from the university's IRB about what
to do in the unlikely event that the Coroner's Office requested cooperation. He
was informed that there was 'no statuary obligation to report criminal activity,' and thus accepted full responsibility for any decision he would make
(quoted in Blomley and Davis 1998).

When he was subpoenaed, his former university refused to offer assistance,
on grounds that 'in cases where it can be foreseen that the researcher may not
legally be in a position to ensure confidentiality to their subjects, these
researchers must be required to prove only limited confidentiality' (quoted in
Lowman and Palys 2000, p. 4). They offered limited financial support only, on
compassionate grounds.

Ogden felt abandoned and believed that SFU was unwilling to protect his
academic freedom as a researcher. Therefore, after successfully having
defended his case before the Court, he filed a lawsuit against the university,
claiming they had a contractual obligation to support his ethical stand and to
reimburse his legal fees.

Though Ogden lost that case, following Bloomley and Davis' 1998 review
of it, the university belatedly accepted responsibility and reimbursed his legal
fees and lost wages and send him a letter of apology, promising to assist
researchers in the future who may find themselves in the position of having to
challenge a subpoena (see Lowman and Palys 2000, for a discussion of the
case). Ogden later became a world leader in suicide research, but his work
remained controversial.

7.5.5   *Waiver of Confidentiality*

Finally, we consider cases where participants themselves wish to be identified or wish to waive their right to confidentiality. Technically these would not be breaches of confidentiality, but rather waivers of confidentiality agreements.

The number of cases in which participants waive confidentiality and/or in which IRBs agree to such a design are uncommon. In certain types of research, however, waivers of confidentiality are the rule rather than the exception. In Participatory Action Research (PAR), participants agree to be 'collaborators' of the researchers, not 'research subjects.' They will not merely be 'interviewees' or 'respondents' of the researcher, but actively engaged in the research process itself, defining together with the researcher the research question and research set up. This means to a degree, the roles of researcher and participant roles blur. And as much as this is the case, there is good reason to give 'special concerns regarding the need to protect confidentiality' say Khanlou and Peter (2005, p. 2338), although that does not necessarily imply lifting it. It means that participants themselves decide how they be involved and define their involvement.

There may be another reason for participants to give confidentiality a second thought. Vainio (2013, p. 689) examined an example in which a researcher conducted a study of an organization, and the individual who developed the organization insisted they be mentioned by name in the report (in the hopes of profiting from it). Here, waiving confidentiality borders on a conflict of interest (see Chap. 8).

7.6   Conclusions

7.6.1   *Summary*

Confidentiality stands as a core tenant of scientific research ethics. Few issues matter more than allowing the participant control over which information they wish to share. The most important procedure in this aspect of research is to secure the *informed consent form*, which formalizes a confidentiality agreement between the researcher and participant. Also, various data points, or *identifiers*, that allow for the re-identification of participants, are important for researchers to understand, as are the techniques to anonymize data, though none offer waterproof guarantee against re-identification. Furthermore, we noted that anonymization in qualitative and quantitative research differs greatly. Finally, *breaches of confidentiality* were discussed, including which forms are justifiable (or excusable), and which are not.

7.6.2   Discussion

Two obstacles regarding confidentiality remain. The first regards the availability of information, and the growing capacity to combine information on a large-scale is making it increasingly difficult to guarantee anonymity. The second is that data protection regulations are still evolving, and the way these regulations coalesce may significantly influence future research agendas.

These two issues – protection of participants' privacy and their autonomy, and evolving data protection regulation – comprise an underlying dilemma: how do you ensure academic freedom while at the same time making sure that everything is done (morally and legally) to protect confidentiality?

Case Study: Too Much Information? A Case Study on Maintaining Confidentiality

The case outlined below highlights some of the difficulties of maintaining scientific standards while simultaneously offering confidentiality, specifically when researching a highly sensitive subject. The following details derive from a group of master's students and their supervisor who were engaged in a field study examining the risks related to sexual and reproductive health (SRH) in a country considered conservative in several respects. Notably in this country, it is a cultural taboo to speak publicly about SRH issues, and accessibility to SRH services remain quite limited as well.

State officials in this country admit that a lack of knowledge on SRH can result in risky sexual behavior and unintended pregnancies, and that these in turn contribute to high rates of sexually transmitted diseases and increased maternal mortality due to (illegal) abortions. While it seems clear that this would justify setting up SRH facilities, a clear government policy on the matter was still lacking, and the emphasis was on countering maternal morality rather than communicating knowledge.

However, the government did allow a network of private SRH care professionals to collaborate with international agencies and Non-Governmental Organizations (NGOs) to initiate a project aimed at filling this gap. Such a project could increase the prevalence of SRH facilities, offering affordable, accessible, quality services which, if successful, could increase awareness and knowledge of SRH, all with the desired outcome of behavioral change.

This project became the focus of the researchers. The project leader granted the students permission to interview key project members and stakeholders. The aim of the study was not to evaluate the offerings of the project as such, but to 'assess the potential indicators that determine success of the project as a case study.'

Prior to beginning their research, the master's students sought and received ethical approval from the local IRB, as well as from their own institutional ethical review board. Due to the sensitivity of the project, it was agreed that the interviewees, the stakeholders, and the organization itself would remain anonymous, and all identifying information would be removed. All participants received an 'informed consent' agreement fully detailing the aims of the study. The agreement also contained a privacy statement that promised full confidentiality. All interviews were recorded, transcribed, and subsequently anonymized.

During the first few weeks of research, interviews were conducted on the participant's expectations, thoughts, and doubts surrounding the project. Many respondents demonstrated an acute awareness of the sensitivities regarding sexual and reproductive health. One stakeholder noted how 'increasing conservatism makes talking about SRH difficult,' and believed that professionals would be 'nervous raising these issues.'

In concluding their research, the master's students stressed the importance of the project for the community. They argued that although it touched upon sensitive issues, the project was neither illegal nor in violation of any state regulations. In order to make the project sustainable, it was recommended that 'partnerships between public and private sector need to be further developed,' and that perhaps 'business experts could be involved to establish a sustainable SRH service model.'

When a draft was presented to the SRH project leader, the students received word that there were still concerns about the 'potential harm' of their research. The students were told that they should consider removing all identifying information about the project from their report. While the project leader admitted that an ethical clearance had been issued by the local IRB earlier, and that promises of confidentiality had been met by the students, drawing attention to the sensitive issue of SRH services in a conservative culture with a taboo on sexual matters could have unforeseen and even adverse consequences for those involved, if not immediately, then perhaps in the future. Therefore, all names of the participants were either to be removed or anonymized, and any references to the actual project be omitted. Additionally, the report was to only be made public if it did not include a description of the SRH project.

This posed a dilemma for the students and their advisor. Firstly, it would be difficult to ensure the quality of their theses without describing the project being studied. Secondly, because their institution required that any master thesis project be submitted and subsequently archived at an institutional repository, it would therefore be made public and open for anyone to inspect in accordance with the scientific demand of transparency. Under the circumstances, it did not seem possible to fulfil participant requests for confidentiality *and* submit a Master's thesis in accordance with university requirements.

In practice, the requirement 'not to publish the report with the description of the project' would imply that the students could not formally finish their research project and thus not conclude their studies. The student's supervisor thereupon arranged for an exception to be made in this case, allowing the report to be archived without the possibility to inspect it, in turn effectively annulling the scientific merits of the study.

When we (the authors of this book) asked to report on this case, the students' supervisor allowed us access to relevant documents, including correspondence with various stakeholders, on the provision that any and all identifying information be removed, including the names of the Master's students, the supervisor, and the SRH project leader, as well the name of the country where the research took place. After having completed our description, we destroyed all documents in our possession pertaining to this case were. Then we asked the supervisor and students involved to review this reconstruction, to see if they approved of it.

Assignment

Consider the nuances of this case. What efforts have the different parties (authors of this case study, project leader, supervisor, and students) pursued to ensure confidentiality? Do you believe these measures were enough? Can you think of another outcome that could have been possible had other steps been taken, and if so, what would you recommend? Discuss the case (and its possible alternative scenarios) in class, and answer the following questions:

1. Is it even possible to research 'sensitive issues' such as sexual and reproductive health, in a conservative culture without endangering the parties involved?
2. If so, what measures should be taken to ensure complete anonymity?
3. How, in the present situation, could a scientist still report on this project safely without giving up on academic freedom?

Suggested Reading

We recommend Sissela Bok's 1983 classic *The Limits of Confidentiality* and the chapter on confidentiality by Slowther and Kleinman (2008) for further orientation on the subject. We also recommend the chapter by Macnish (2020) in the *Handbook of Research Ethics and Scientific Integrity* (Iphofen, ed., 2020) for a discussion on the challenges inherent to privacy. Furthermore, we point to Manson and O'Neill (2007) for an extensive discussion on informed consent (though it is mainly focused on the medical sciences).

References

Amiri, F., Yazdani, N., Shakery, A., & Chinaei, A. H. (2016). Hierarchical anonymization algorithms against background knowledge attack in data releasing. *Knowledge-Based Systems, 101*, 71–89. https://doi.org/10.1016/j.knosys.2016.03.004.
Behnke, S. (2014, April). Disclosing confidential information. *Monitor on Psychology, 45*(4). http://www.apa.org/monitor/2014/04/disclosing-information.

Bersoff, D. N. (2014). Protecting victims of violent patients while protecting confidentiality. *American Psychologist, 69*(5), 461–467. https://doi.org/10.1037/a0037198.

Bjarnason, T., & Adalbjarnardottir, S. (2000). Anonymity and confidentiality in school surveys on alcohol, tobacco, and cannabis use. *Journal of Drug Issues, 30*(2), 335–343. https://doi.org/10.1177/002204260003000206.

Blomley, N., & Davis, S. (1998). *Russel Ogden decision review*. Online: SFU President's Homepage, http://www.sfu.ca/pres/OgdenReview.htm (date accessed: 12 Mar 2020).

Bok, S. (1983). The limits of confidentiality. *The Hastings Center Report, 13*(1), 24–31. https://www.jstor.org/stable/3561549.

Buchmann, E., Böhm, K., Burghardt, T., et al. (2013). Re-identification of smart meter data. *Personal and Ubiquitous Computing, 17*, 653–662. https://doi.org/10.1007/s00779-012-0513-6

Ciriani, V., di Vimercati, S. D. C., Foresti, S., & Samarati, P. (2008). K-anonymous data mining: A survey. In C. C. Aggarwal & P. S. Yu (Eds.), *Privacy-preserving data mining. Advances in database systems, vol. 34* (pp. 105–136). Berlin: Springer. https://doi.org/10.1007/978-0-387-70992-5_5.

Dawson, P. (2014). Our anonymous participants are not always anonymous: Is this a problem? *British Journal of Educational Technology, 45*(3), 428–437. https://doi.org/10.1111/bjet.12144.

Domingo-Ferrer, J., & Torra, V. (2005). Ordinal, continuous and heterogeneous k-anonymity through microaggregation. *Data Mining and Knowledge Discovery, 11*(2), 195–212. https://doi.org/10.1007/s10618-005-0007-5.

Duncan, R. E., Hall, A. C., & Knowles, A. (2015). Ethical dilemmas of confidentiality with adolescent clients: Case studies from psychologists. *Ethics & Behavior, 25*(3), 197–221. https://doi.org/10.1080/10508422.2014.923314.

El Emam, K., Jonker, E., Arbuckle, L., & Malin, B. (2011). A systematic review of re-identification attacks on health data. *PLoS One, 6*(12), e28071. https://doi.org/10.1371/journal.pone.0028071.

Geraghthy, R. (2016). *Anonymisation and social research*. Anonymising Reserch Data Workshop, University College Dublin, 22 June 2016. www.slideshare.net/ISSDA/anonymisation-and-social-research

Giordano, J., O'Reilly, M., Taylor, H., & Dogra, N. (2007). Confidentiality and autonomy: The challenge(s) of offering research participants a choice of disclosing their identity. *Qualitative Health Research, 17*(2), 264–275. https://doi.org/10.1177/1049732306297884.

Herbert, P. B. (2002). The duty to warn: A reconsideration and critique. *Journal of the American Academy of Psychiatry and the Law Online, 30*(3), 417–424. Retrieved from https://pdfs.semanticscholar.org/5a4c/b550a640d165ec49c5a922291961c278aee6.pdf.

Hook, M. K., & Cleveland, J. L. (1999). To tell or not to tell: Breaching confidentiality with clients with HIV and AIDS. *Ethics & Behavior, 9*(4), 365–381. https://doi.org/10.1207/s15327019eb0904_6.

Iphofen, R. (Ed.). (2020). *Handbook of research ethics and scientific integrity*. Cham: Springer.

Irwin, S. (2013). Qualitative secondary data analysis: Ethics, epistemology and context. *Progress in Development Studies, 13*(4), 295–306. https://doi.org/10.1177/1464993413490479.

Israel, M. (2014). *Research ethics and integrity for social scientists* (2nd ed.). London: Sage.

Kaiser, K. (2009). Protecting respondent confidentiality in qualitative research. *Qualitative Health Research, 19*(11), 1632–1641. https://doi.org/10.1177/1049732309350879.

Kelly, A. (2009). In defence of anonymity: Re-joining the criticism. *British Educational Research Journal, 35*(3), 431–445. https://doi.org/10.1080/01411920802044438.

Khnalou, N., & Peter, E. (2005). Participatory action research: Considerations for ethical review. *Social Science & Medicine, 60*(10), 2333–2340. https://doi.org/10.1016/j.socscimed.2004.10.004.

Kotch, J. B. (2000). Ethical issues in longitudinal child maltreatment research. *Journal of Interpersonal Violence, 15*(7), 696–709.

Kumpošt, M., & Matyáš, V. (2009). User profiling and re-identification: Case of university-wide network analysis. In S. Fischer-Hübner, C. Lambrinoudakis, & G. R. Pernul (Eds.),

Trust, privacy and security in digital business (pp. 1–11). Berlin: Springer. https://doi. org/10.1007/978-3-642-03748-1_1.

LaFrance, J., & Bull, C. C. (2009). Research ourselves back to life. Taking control on the research agenda in Indian country. In D. M. Mertens & P. E. Ginsberg (Eds.), *The handbook of social research ethics* (pp. 135–149). London: Sage. https://doi.org/10.4135/9781483348971.n9.

Lelkes, Y., Krosnick, J. A., Marx, D. M., Judd, C. M., & Park, B. (2012). Complete anonymity compromises the accuracy of self-reports. *Journal of Experimental Social Psychology, 48*(6), 1291–1299. https://doi.org/10.1016/j.jesp.2012.07.002.

Lowman, J., & Palys, T. (2000). Ethics and institutional conflict of interest: The research confidentiality controversy at Simon Fraser University. *Sociological Practice: A Journal of Clinical and Applied Sociology, 2*(4), 245–255. https://doi.org/10.1023/A:1026589415488.

Lubarsky, B. (2017). Re-identification of "anonymized data". *Georgetown Law Technology Review, 202*, 202–213. https://perma.cc/86RR-JUFT.

Macnish, K. (2020). Privacy in research ethics. In R. Iphofen (Ed.), *Handbook of research ethics and scientific integrity* (pp. 233–249). Cham: Springer. https://doi.org/10.1007/978-3-319-76040-7.

Manson, N. C., & O'Neill, O. (2007). *Rethinking informed consent.* Cambridge: Cambridge University Press.

Martin, D. J., Kifer, D., Machanavajjhala, A., Gehrke, J., & Halpern, J. Y. (2007, April). Worst-case background knowledge for privacy-preserving data publishing. In *2007 IEEE 23rd international conference on data engineering* (pp. 126–135). Piscataway: IEEE. https://doi. org/10.1109/ICDE.2007.367858.

Munson, R. (2008). *Intervention and reflection: Basic issues in medical ethics* (8th ed.). Belmont, CA: Thomson Wadsworth.

Narayanan, A., & Shmatikov, V. (2009). De-anonymizing social networks. In *2009 30th IEEE symposium on security and privacy* (pp. 173–187). Los Alamitos: IEEE. https://doi.org/10.1109/ SP.2009.22.

Ohm, P. (2010). Broken promises of privacy: Responding to the surprising failure of anonymization. *UCLA Law Review, 57*(6), 1701–1778. https://ssrn.com/abstract=1450006.

Patel, D. (2016). Research data management: A conceptual framework. *Library Review, 65*(4/5), 226–241. https://doi.org/10.1108/LR-01-2016-0001.

Ramachandran, A., Singh, L., Porter, E., & Nagle, F. (2012). Exploring re-identification risks in public domains. In *2012 tenth annual international conference on privacy, security and trust* (pp. 35–42). Paris: IEEE. https://doi.org/10.1109/PST.2012.6297917.

Rhoen, M. H. C. (2019). *Big data, big risks, big power shifts: Evaluating the general data protection regulation as an instrument of risk control and power redistribution in the context of big data (doss.).* Leiden: Leiden University. https://openaccess.leidenuniv.nl/handle/1887/77748.

Samarati, P., & Sweeney, L. (1998). *Protecting privacy when disclosing information: k-anonymity and its enforcement through generalization and suppression.* Technical report, SRI International. Retrieved from: https://epic.org/privacy/reidentification/Samarati_Sweeney_paper.pdf

Scarce, R. (1995). Scholarly ethics and courtroom antics: Where researchers stand in the eyes of the law. *The American Sociologist, 26*(1), 87–112. https://doi.org/10.1007/BF02692012\.

Scott, R. C. (1995). Anonymity in applied communication research: Tension between IRBs, researchers, and human subjects. *Journal of Applied Communications, 333*, 242–257. https:// doi.org/10.1080/00909880500149445.

Slowther, A., & Kleinman, I. (2008). Confidentiality. In P. A. Singer & A. M. Viens (Eds.), *The Cambridge textbook of bioethics* (pp. 43–50). Cambridge: Cambridge University Press.

Taddei, S., & Contena, B. (2013). Privacy, trust and control: Which relationships with online self-disclosure? *Computers in Human Behavior, 29*(3), 821–826. https://doi.org/10.1016/j. chb.2012.11.022.

Thomson, D., Bzdel, L., Golden-Biddle, K., Reay, T., & Estabrooks, C. A. (2005). Central questions of anonymization: A case study of secondary use of qualitative data. *Forum: Qualitative Social Research, 6*(1), Art. 29, http://nbn-resolving.de/urn:nbn:de:0114-fqs0501297.

Tilley, L., & Woodthorpe, K. (2011). Is it the end for anonymity as we know it? A critical examination of the ethical principle of anonymity in the context of 21st century demands on the qualitative researcher. *Qualitative Research, 11*(2), 197–212. https://doi.org/10.117 7/2F1468794110394073.

Tolich, M. (2004). Internal confidentiality: When confidentiality assurances fail relational informants. *Qualitative Sociology, 27*(1), 101–106. https://doi.org/10.1023/B:QUAS.0000015546 .20441.4a.

Vainio, A. (2013). Beyond research ethics: Anonymity as 'ontology', 'analysis' and 'independence'. *Qualitative Research, 13*(6), 685–698. https://doi.org/10.1177/2F1468794112459669.

Van den Hoonaard, W. C. (2003). Is anonymity an artifact in ethnographic research? *Journal of Academic Ethics, 1*(2), 141–151. https://doi.org/10.1023/B:JAET.0000006919.58804.4c.

Walford, G. (2005). Research ethical guidelines and anonymity. *International Journal of Research & Method in Education, 28*(1), 83–93. https://doi.org/10.1080/01406720500036786.

Weinberg, M. (2002). Biting the hand that feeds you and other feminist dilemmas in fieldwork. In W. C. van den Hoonaard (Ed.), *Walking the tightrope: Ethical issues for qualitative researchers* (pp. 79–94). Toronto: University of Toronto Press.

Whelan, T. J. (2007, October). *Anonymity and confidentiality: Do survey respondents know the difference? Poster presented at the 30th annual meeting of the Society of Southeastern Social Psychologists*. Durham, NC.

Wiles, R., Charles, V., Crow, G., & Heath, S. (2006). Researching researchers: Lessons for research ethics. *Qualitative Research, 6*(3), 283–299. https://doi.org/10.1177/2F1468794106065004.

Williams, G., & Pigeot, I. (2017). Consent and confidentiality in the light of recent demands for data sharing. *Biometrical Journal, 59*(2), 240–250. https://doi.org/10.1002/bimj.201500044.

Zang, H., & Bolot, J. (2014). Anonymization of location data does not work: A large-scale measurement study. *2014 IEEE International Conference on pervasive computing and communication workshops* (PERCOM WORKSHOPS), Budapest, Hungary, 24–28 March 2014. https:// doi.org/10.1145/2030613.2030630.

Zhou, B., Pei, J., & Luk, W. S. (2008). A brief survey on anonymization techniques for privacy preserving publishing of social network data. *ACM Sigkdd Explorations Newsletter, 10*(2), 12–22. https://doi.org/10.1145/1540276.1540279.

Zimmer, M. (2010). 'But the data is already public': On the ethics of research in Facebook. *Ethics and Information Technology, 12*(4), 313–325. https://doi.org/10.1007/s10676-010-9227-5.

Chapter 8
Conflicts of Interest

Contents

Electronic Supplementary Material: The online version of this chapter (https://doi.org/10.1007/978-3-030-48415-6_8) contains supplementary material, which is available to authorized users.

After Reading This Chapter, You Will:

- Have an understanding of how modern scientific research is influenced by economic interests
- Be able to identify conflicts of interest and conflicts of values
- Be capable of recognizing competing interests
- Understand conflicts of ownership
- Know how various conflicts are resolved

Keywords Academic capitalism · Accountability · Autonomy · Bribery · Disclosure (of interest) · Globalization · Patenting · Conflict of interest · Competing interests · Funding bias · Objectivity · Reporting bias · Research alliance · Trustworthiness · Valorization

8.1   Introduction

8.1.1   Science in a Global Society

In the last quarter of the twentieth century, scientists began noticing that the social and political landscape was changing rather rapidly. This change was beginning to impact the organization of science too, including the way it is financed. Also, the process of production and dissemination of scientific knowledge transformed, calling into question the very values of science.

Two related trends brought about these changes. One was the onset of 'globalization,' a series of processes that caused the world's economies, cultures, and populations to become more interdependent, simultaneously intensifying interactions among people, companies, and governments on a global scale (Bartelson 2000; Hoekman 2012; Mosbah-Natanson and Gingras 2013). The other is the introduction of neoliberalism in politics (resurgent of nineteenth century economic laissez-faire ideas and free market capitalism). This soon found its way into university administration (see Fine and Saad-Filho 2017, for a discussion of neo-liberalism, with its internal contradictions, tensions, and sources of dynamics).

During this time period, more emphasis was placed on 'output,' 'quality control,' and 'cost-efficiency,' on science's 'added value' to society. Many scientists saw new opportunities in these developments and began to establish new collaborative relationships, especially with commercial parties. More space for entrepreneurial activity was created and universities began to enhance their social impact on society. Incubators (start-up organizations and businesses involved with advanced technology) found their way onto college campuses, and universities flourished as creative business environments.

At the same time, new challenges and tensions emerged, having to do with the growing competition over dwindling public funding and a 'pressure to produce.' A trend towards *valorization of research* was also observed, which meant that the economic value of scientific activities began to predominate, not its intrinsic value (see Box 8.1). Some even began to speak of 'academic capitalism' (Slaughter and Roades 2004 (Fig. 8.1).

Box 8.1: Valorization and Academic Careers

'Universities, Academic Careers, and the Valorization of "Shiny Things"' is the title of a 2016 paper by Joseph Hermanowicz, professor of sociology at the University of Georgia. Hermanowicz researched the careers of some 60 physicists employed at universities across the U.S. between the mid-1990s and mid-2000s, when neoliberalism 'enabled academic capitalism to flourish with its attendant effects in privatization and marketization' (2016, pp. 303–4). How did this affect their careers and aspirations? Here are a two notable changes Hermanowicz found:

Publish and Perish Institutions began to favor research which brought about a 'press for productivity.' By the time the older generation (who entered university before 1970) became tenured professors, they had published on average of about 20 papers. The younger generation, in comparison, published over 40 by the time they became professors. Doubling the number of publications in roughly the same time span meant an increased pressure to publish.

Disappointment Greater output did not result in more job satisfaction or commitment. On the contrary, actually. Many scientists were disappointed, even bitter, about the insistence on greater output. Although it differs whether one works at an 'elite' institute, which is largely committed to research, or one that emphasizes (mass) teaching and services, job satisfaction and work attitude seems to be affected negatively by valorization throughout different cohorts. 'Professors increasingly understand themselves and their work in terms of free agency, geared to a market, and interested especially if not exclusively in themselves' (Hermanowicz 2016, p. 324).

Students' Change of Focus Adding to this picture, Saunders (2007) found a change in the focus of students during the same period. In his findings, the majority of students switched from being intrinsically motivated, having an interest in developing a 'meaningful philosophy of life', to being extrinsically motivated, having an interest in being 'well off.' Many students now agree with the statement that 'the chief benefit of a college education is to increase one's earning power.' Students today are more competitive, less interested in liberal arts and sciences, and seem to subscribe to a neoliberal agenda.

Fig. 8.1 Age of academic capitalism

It has been argued that a growing emphasis on protection of intellectual property, and a strong preference for profit, has corroding effects on some of the core values in science, such as openness, transparency, and autonomy. In turn, these invited a host of unwanted behaviors in research, ranging from fraud to questionable decisions, which we will discuss in greater detail below (see Washburn 2005; Healy 2002; Greenberg 2009).

In this chapter, we explore the links between the three parties involved in these developments, namely science, private industry, and government. Traditionally, these three parties had clear, distinct roles. Sciences' task was restricted to research and investigation; industry focused on development and production; and government in creating and enforcing regulation. In the age of *academic capitalism* however, the boundaries between these roles gradually dissolved, resulting in a number of emergent conflicts. It is these various conflicts that we will discuss in this chapter.

8.1.2 Interests at Stake

New forms of collaboration between researchers and third parties bring special interests into play which hitherto did not have such importance. How do the interests of collaborative parties influence a researcher's judgement?

Consider the case of a researcher who was commissioned by a political party to investigate a certain social question. Would it matter if the researcher has strong political convictions themselves? Would it make a difference if these convictions align, or differ, with those of the employing political party? Or consider a researcher investigating the effectiveness of certain policies implemented by an organization. Would it matter if these researchers are also being paid to provide training to members of that same organization? Would that affect their assessment of the policies being examined?

Most would agree that these situations contain an element of suspicion. How do we know the researcher in question is *not* influenced by certain external interests, either financial or otherwise?

When there is reason to believe that a researcher's judgement may be compromised by certain interests related to their ties with other parties, we commonly speak of a *conflict of interest* (see Davis 2001). Some authors, such a Aubert (1963), and more recently Bero and Grundy (2016) and Wiersma et al. (2017), have suggested that we distinguish 'conflicts of interest proper' (having to do with financial motives only) from 'conflicts of values' (having to do with morality and belief systems), which are both structured in fundamentally different ways and require divergent strategies to resolve.

We accept the suggestion to separate financially induced *conflicts of interest* (COI), which mostly appear in privately funded research, from *conflicts of values* (COV), which arise out of various types of partnerships with thirds parties (nonprofit organizations, interest groups, civil society organizations, and others). We furthermore propose to highlight two additional categories, namely *competing interests*, which are more often found in publicly funded research and do not directly impact judgement calls, and *conflicts of ownership*, which have to do with questions of copyrights and patents.

In the coming sections, we discuss how these various conflicts impact scientific norms in different ways (see Fig. 8.2 for an overview).

	Cooperative relationship with whom?	Which interests involved?	Which values at stake?
Conflict of Interest	Funding agencies, Industries, Lobbyists	Financial Economic	Objectivity
Conflict of Values	Partners, Associates, Research Allies	Personal Situational Disciplinary	Reliability
Competing Interests	Governmental Bodies Interest Groups, Clients, Civilians	Ideological Cultural	Respect of Others and Carefulness
Conflict of Ownership	Owners of Networks and Data, Copyright Holders	Judicial	Autonomy and Accountability

Fig. 8.2 Types of conflicts in collaborative relationships

8.2 Conflicts of Interest

8.2.1 What Is a Conflict of Interest?

The classical situation in which a researcher's decision-making may be compromised because of certain financial interests is called a *conflict of interest* (COI). Conflicts of interest are more common in the bio-medical and pharmaceutical sciences, where large financial gains are at stake, and the development of new medication is a costly affair. In the social sciences, financial conflicts of interest do exist but the temptations differ from those of the bio-medical and pharmaceutical science.

Let's start with an example from the pharmaceutical sciences. Resnik (1998) cites a classic case of a scientist who researched the effects of a certain medication on the alleviation of common cold symptoms. The scientist also owned stock in a company that produced the same medication he was researching (a tablet of zinc lozenges). When their findings showed a positive result, the company's stock soared, from which the researcher benefited. This raised a serious question: Was the researcher's scientific judgement being influenced by the expectation of a financial profit?

In the social sciences, direct financial gains are rarer. Rather, the problem lies in indirect gains, having to do with the formation of dependency on the research itself. Soudijn (2012) quotes the case of a Dutch psychologist, who set up a project offering help to clients suffering from phobias. The clients received free treatment (in the form of experimental therapy, given by his students) on the condition that they agreed to participate in the research project. Thus, the clients became reliant on the research as a means of free therapy. These dependency relationships obfuscate the research project to the point that by today's standards, the data would no longer be considered valid, and although the research participants did not profit from the research financially, financial gains (free therapy for the client) posed a COI in this case.

Whether these influences actually impair a researcher's judgement is not of importance in our understanding of a COI. It is the *potential* to cloud or impair judgement that defines the problem.

In any conflict of interest, *objectivity* as one of sciences' key values is at stake:

• How do I know your conclusions are not biased?
• How can I trust your judgement?

In the coming sections, we discuss cases from within the social sciences where differing financial interests were at stake to differing degrees (Box 8.2). Note that not every situation with financial interests at stake automatically leads to a conflict of interest. Furthermore, it can be difficult to establish whether a researcher acts in bad faith or not.

Box 8.2: Funding Bias

Often regarded as a specific form of COI, the term *funding bias* indicates the tendency found in scientific studies to support the interests of the study's financial sponsor. Funding bias is a well-documented effect (see Krimsky 2012). A study by Turner and Spilich (1997) well illustrates this conclusion. The authors compared 91 papers written about the effects that tobacco and nicotine have on cognitive performance. The authors differentiated between industry sponsored and independent studies. Both types of studies reported on the positive effects of tobacco and nicotine use (indications that tobacco enhances cognitive performance). However, all non-sponsored studies also reported negative effects (indications that tobacco does *not* enhance cognitive functioning), while only one sponsored study did so.

While non-sponsored studies thus present more balanced results, there can be disadvantages to not having external sponsorship too. It can mean that the researchers have not succeeded in developing links to the field of expertise, and that their work is 'sterile.' Martens et al. (2016), reporting on the field of counseling research, noted that 'sponsored research has the potential to advance the professional goals of a field by allowing researchers to conduct studies of a scope and magnitude that they would not be able to do without external funding' (p. 460).

8.2.2   Promotion Bonus

In a commercialized academic climate at Tilburg University, a sudden increase in the number of successfully completed PhD dissertations in the field of the humanities and the social sciences was observed a few years ago. They were carried out under the supervision of a small number of professors. Suspicion arose that this increase had to do with financial interests, as each dissertation produced financial yields, and some of the revenue was being paid to the supervisors directly (they received a *promotion bonus*).

A complaint was issued. The commission that investigated the cases found that while many of the dissertations were rather weak and held very little scientific value, there was no evidence that fraud was being committed. Did this cancel out the possibility of a conflict of interest? Not entirely. The commission strongly recommended that the promotion incentives be abolished to preemptively avoid any possible suggestions of COIs. Tilburg University complied with this advice and no longer pays these incentives out to supervisors (case reported in Universe, September 20th, 2018).

8.2.3   *Financial Ties*

In a report on the financial links between panel members who were asked to contribute to the development of diagnostic criteria for the *Diagnostic and Statistical Manual of Mental Disorders* (DSM) and the pharmaceutical industry, Cosgrave et al. (2006) found notable examples of said links. More than half of the panel members queried had financial ties with a pharmaceutical company; 40% of them had consulting income; 29% served on a speakers bureau; and 25% received other professional fees (2006, p. 156).

Does the presence of financial ties imply that these panel members can't be trusted? Not necessarily. Cosgrave et al. (2006) argue that receiving financial support does not automatically disqualify members, but it *does* pose a conflict of interest. 'The public and mental health professionals have a right to know about these financial ties, because pharmaceutical companies have a vested interest in what mental disorders are included in the DSM' (p. 159) (Box 8.3).

Box 8.3: Mutual Favors: A Dilemma

Consider and discuss in groups of three or four students the following case and select the best course of action:

Jacky is a student enrolled in the same course as you. When you are assigned to write a paper on a certain subject together, Jacky makes you the following offer. If you do the work this time and put her name on the paper, she will do the same for you next time. You are both pressured for time, but Jacky has been particularly overloaded and could use some help. The syllabus allows students to team up on papers – but not in this manner.

How do you respond? Choose the option you prefer and present your rationale to your group.

1. You allow Jacky co-authorship on the paper but you decline her offer to return the favor.
2. You accept the offer, on the condition that you both commit to a critical peer review of each paper.
3. You ask for advice from a professor outside the course, who also happens to know Jacky.
4. You decline the offer and report the unethical behavior to the professor.

(Case adapted after the Erasmus University *Dilemma Game*)

8.2.4 Paid Consultancies

Do financial interests resulting from paid lectures and expert consulting pose a conflict of interest? This question was raised in the case of Jean Twengy, a psychologist at San Diego State University. Twengy studied people born after the mid-1990s, which she dubbed 'iGen' (after the iPhones this generation grew up with). In 2010, she started a business called 'iGen Consulting'. As a consultant, she gave paid lectures and advised large commercial corporations, acquiring a considerable side income. In her academic papers, she did not mention this revenue.

Some argue this poses a conflict of interest. Psychology journals require that personal fees be declared, and since Twengy had not done so, her side activities qualify as a COI by default, they said. Others however, including psychologist Steven Pinker from Harvard University, argue they do not, because 'these activities do not provide incentives to make certain judgement calls.' Pinker argues that Twengy's case does not compare to 'evaluating a drug produced by a company in which one holds stock' (case discussed in Chivers 2019). It is true that these financial interests did not relate directly to the researcher's field of inquiry. However, they do so indirectly, since 'iGen' is the basis of both her research and her advisory work.

8.2.5 Gifts

In pursuit of any research project, accepting gifts or 'gratuities' should be avoided or kept to the bare minimum (such that the gifts represent no substantial value). At the very least, it can leave the impression that favorable opinions can be bought. Accepting something of value without the other party expecting something in return constitutes a conflict of interest because of the suggestion of favoritism. And if something of value is accepted and the other party *does* expect something in return, this is even worse. It's called *bribery*.

A bibliography study by Volochinsky et al. (2018) concluded that conflicts of interest produced by gift giving result in an 'undeniable dependence on relationships that implicitly lead to reciprocity and deprive the professionals from the necessary autonomy and impartiality' (p. 101). But there is another side to this story as well.

Anthropologists, who have long emphasized the importance of gift giving in human societies, argue that a strict view may be too narrow (Mauss 1970; Sherry 1983). Some noted that specifically in Asia, gift giving is part of an 'exchange system,' and non-participation will be considered rude and disruptive of social relationships. Respondents in the research of Zhuang and Tsang (2008) were found to have 'different ethical evaluations of different marketing practices.'

Gift giving in a natural setting poses an ethical dilemma for researchers: either go along with the practice of gift giving and run the risk of a possible conflict of interest, or stay within the strict boundaries of research ethics and accept that certain research goals may not be achieved as a result. (Box 8.4)

Box 8.4: Big Pharma
The expression 'big pharma' refers to the 'pharmaceutical industry,' the private sector that invests billions into medical research and services. Pharmaceutical companies employ various (often aggressive) means to protect their investments, and for this reason the term 'big pharma' is loaded with a fair amount of distaste or even mistrust (see for example: Angell 2004, *The Truth About the Drug Companies: How They Deceive Us and What to do About It*).

One such critic is David Resnik, whose 2007 book *The Price of Truth* provides ample evidence of how academic values such as objectivity, truthfulness, and openness have been corrupted by financial interests. Another critic, Ben Goldacre (2013), wrote critically on the naive view that the public has of medical research and doctors, and how they often fail to acknowledge just how much financial interests shape the medical field.

Johan Hari (2018) writes in a similar vein, in his case focusing on the field of clinical psychology, which he argues has become equally dominated by the pharmaceutical industry. Drug companies, he claims, are aggressively advertising an image of depression as a 'chemical imbalance' in the brain, with the sole intent of selling drugs to the clinically depressed. They fund scores of studies, 'cherry pick' those that corroborate the positive effects of the drugs they produce, and 'kill' the ones that challenge their narrative.

8.3   Conflicts of Values

8.3.1   What are Conflicts of Values?

Similar to *conflicts of interest* (COI), *conflicts of values* (COV) compromise the researcher's judgement, or decision-making. The main difference being that the source of commitment in a conflict of values is not financial in nature.

Thus, if a researcher has close working relationships with research affiliates, is on personal terms, or has a dependent relationship with them, conflicts of values may arise. For example, if a researcher is asked to assess the work of one of their personal associates, it is difficult to avoid the suspicion that their judgement is influenced by this relationship.

In a straightforward example of a conflict of values, a researcher from Shanghai University, acting as the handling editor of a research journal, was asked to process a recently received manuscript. After receiving a positive review from an anonymous reviewer, the paper was published. Not long thereafter, it was discovered that the handling editor had co-authored several papers with the author of the recently published paper. It was also revealed the author was a former PhD student of the handling editor. The paper was subsequently retracted.

Although it is entirely possible that the relationship between the author and han-dling editor played no factor in the matter, for the sake of scientific disinterested-ness, journals must avoid any suggestion of 'favoritism.' And indeed, the journal changed its review policy after the fact, such that no future editor would be allowed to deal with manuscripts submitted by colleagues or 'research alliances' (partners involved in the same research). (case derived from Retraction Watch, 'We would now catch' this conflict of interest, entry September 8th, 2017).

In our understanding of conflicts of values, the values of *reliability* are at stake:

- How do I know your assessment is fair?
- How can I trust your valuation is not prejudiced?

Below we discuss a few cases in which conflicting values factored into a researcher's ability to weigh the situation fairly or justly.

8.3.2 Allegiance Effect

Maj (2008) discusses a phenomenon known in clinical research as the *allegiance effect*, the propensity of researchers to favor the school of thought they belong to. This effect plays out in ways similar to how financial conflicts of interest impact drug trials. Maj (2008, p. 91) found that the allegiance effect occurs through the 'selection of a less effective intervention to compare with the researcher's favorite treatment; unskillful use of the comparison treatment; focusing on data favoring the preferred treatment in study reports; and failure to publish negative data.' In short, researchers purposefully restrain themselves to a specific, desirable body of knowledge.

8.3.3 Disciplinary Bias?

Scientists often feel that grant application review processes are skewed. Some believe that certain disciplines are favored over others, or that certain approaches are more likely to get selected over others. They liken it to rivalries and disciplinary prejudices in peer reviewing. Should this be the case, then conflicts of values, rather than conflicts of interest, are at play, because there are no direct monetary gains to be expected by the reviewers.

A team of French researchers interviewed 98 scientists involved in grant review processes (either as a reviewer or as an applicant) and found that many did in fact believe that *disciplinary bias* played a role in the process, although they admitted that this was impossible to prove. Nonetheless, many still felt that certain disci-plines strongly support their research topics, and that favoritism was almost self-evident. One reviewer even admitted to being 'much more lenient […] with the people we know' (Abdoul et al. 2012, e35247).

8.4   Competing Interests

8.4.1   What are Competing Interests?

Any situation where collaborating researchers have unaligned intentions or desired outcomes is a situation of competing interests. Throughout the research process, different perspectives and sometimes incompatible expectations are brought into the play.

This can be seen, for example, in commissioned research, when a researcher is asked to provide advice in a matter that interests multiple stakeholders, such as governmental bodies, individual clients, and industry professionals. These stakeholders may envision contrary approaches to and focus of the research; they may bring in different values, risk assessment strategies, and expectations.

To help differentiate conflicts of interest and conflicts of values from competing interests, note that the former is in play when a researcher's judgement is compromised, but that in the latter it is not the researcher's decision-making that is called into question. Rather, what is at stake is a researcher's ability to reach consensus or agreement with multiple stakeholders. Will the research fulfill the needs of all parties? Will they be able to assess the situation justly?

In our understanding of competing interests, the key value at stake is *carefulness*:

• How do I know that the parties involved are represented fairly?
• How do I know that the stakeholders' interests are weighed properly?

From here, we will discuss cases in which competing interests affected a researcher's ability to weigh a situation fairly or justly.

8.4.2   Wicked Problems

Head and Alford (2015) discuss the difficulties inherent to researching so called *wicked problems* in public policy and management research. Wicked problems are unpredictable and open-ended questions that involves a great number of people and present a major challenge to researchers (for example climate change, or pandemic influenza). Problems become 'wicked' when they elicit the unforeseen consequences of policy interventions.

Most policy researchers acknowledge that wicked problems cannot be solved through the traditional 'engineering approaches.' Instead, researchers must pay attention to the deep-rooted disagreements between stakeholders about the nature, significance, goals, and solutions to these problems. This requires that researchers address 'value perspectives' that allow for certain forms of 'bargaining' in the political marketplace, aimed at reaching a shared understanding, agreed purpose, and mutual trust. The researcher's ethical challenge here is to negotiate a balanced perspective that fairly addresses differing perspectives while remaining open-ended at the same time.

Box 8.5: Sources of Conflicts

Conflicts of Interest (COI)	Conflicts of Values (COV)	Competing Interests (CI)	Conflicts of Ownership (COO)
Employment in a commercial firm	Family connections	Culture	Copyrights
Paid consulting	Research affiliation	Ideology	Patents
Financial incentives	Theoretical allegiances	Religion	Data-ownership
Professional fees	Dependent relationships	Politics	
Stock ownership	Friendships		
Gifts	Publication history		
In-kind support of materials			
Research funding or grant support			

8.4.3   Ideological Interests

An *ideologically informed bias*, as explored by MacCoun (2015), serves as a basis of competing interests. MacCoun argues that in public policy, the ultimate goal (for example, income equality, reproductive rights, welfare entitlements) are not self-evident and in fact are often contested. This can lead to suspicion that a researcher may be biased in their assessment of the outcomes. However, as MacCoun found, these politically bent biases can also be harmful. This can be seen when adversaries who don't like a researcher's findings are granted a greater chance of questioning their opponent's motives when an apparent bias is present (Box 8.5).

8.5   Conflicts of Ownership

8.5.1   What Is a Conflict of Ownership?

There are situations in which the use of findings, data, or other aspects of research is restricted by parties who claim ownership over them. They may be external financers, sponsors, commissioners, commercial parties, or any other entity who has secured rights to intellectual property (the right to own an idea or discovery). These parties may demand that certain restrictions apply were the research to be shared or utilized. If these restrictions contradict scientific norms (for example, refusing open access), we propose to speak of them as *conflicts of ownership* (COO).

In our understanding of conflicts of ownership, there are two key values at stake, namely *autonomy* and *accountability*:

- What level of freedom do researchers have to make their own decisions?
- What restrictions imposed by others apply to their work?

The difference between a conflict of interest and a conflict of ownership lies in the fact that in a COO, it is not the judgement of the researcher that is called into question. Rather, it is the people with whom the researcher is working. Conflicts of ownership differ from competing interests in that the researcher is not at liberty to decide how or what to research, or how to use any subsequent findings.

8.5.2 Patents

Intellectual property is the right to own an idea. It allows the holder control over intellectual property. Patents challenge the communal ideal that science belongs to everybody, and that property right should be kept to a minimum (see Chap. 2 for a discussion of these ideals). Though patents are mostly associated with the natural sciences, the life sciences, and pharmacology, they are issued in the social sciences as well. Social science patents can be seen, for example, psychological testing protocols; biofeedback methods (devices that determine the psychological state of a subject); or educational games, methods, and instruments (see Box 8.6).

8.5.3 Copyrights

Owners of intellectual property have the right to reproduce or disseminate the research data they possess. The rights afforded to them may include the right to withhold publication or to frame research in certain ways. Copyrights very much challenge the communal ideal of science when they are abused.

When sponsors of a study are interested in certain findings only, and they own the copyrights, they can decide to insist on incomplete or partial reporting. The result is referred to as *reporting bias* (under-reporting of unexpected or undesirable results).

Song et al. (2010) found ample evidence of reporting bias in their comparative research of 300 empirical studies, identifying that pressure from research sponsors, instruction from journal editors, and intricacies of the research award system (who is eligible to receive which grants?) play substantial roles in their work and how they report it. 'Clearly, commercial and other competing interests of research sponsors and investigators may influence the profile of dissemination of research findings (Song et al. 2010, p. 81).

Box 8.6: Patenting Psychological Tests: Costs and Benefits
Since the beginning of the twentieth century, social scientists have been searching for instruments that allowed them to measure 'psychological constructs,' such as intelligence, dimensions of personality, emotions, and other 'qualities of the mind.' From the beginning, economic considerations were the primary impetus for psychological assessment. The driving thought was that cheap and cost-effective testing should replace inefficient and costly interviews that would provide the same information (see Yates and Taub 2003).

As a result of this mission, a large number of psychological tests have been developed and are now being used in clinical and developmental psychology, industrial and organizational psychology, forensic psychology, education, and in other applied fields of the social sciences. In these areas, psychological tests serve a variety of purposes, such as in the selection of candidates, assessment of professional abilities, prediction of developmental trajectories, or planning of treatment.

As psychological testing has become standing practice in various occupational areas of Western society, the amount of money invested in these practices has increased accordingly. Psychological assessment has become an extremely profitable business and consequently, the copyright holders of psychological tests (often the publishers) have guarded access to and use of these instrument, which they consider 'trade secrets,' similar to how pharmaceutical industries patent their products.

For a prime example, the publisher Pearson, who owns a number of psychological tests, asserts that 'test questions and answers, manuals and other materials divulging test questions or answers constitute highly confidential, proprietary testing information which Pearson takes every precaution to protect from disclosure beyond what is absolutely necessary for the purpose of administering the test.'

Their products are sold 'only to qualified individuals who are bound by the ethical standards of their profession to protect the integrity of the materials by maintaining the confidentiality of the questions and answers' (www.pearson-clinical.co.uk/information/legal-policies.aspx).

8.6 Resolving Conflicts

8.6.1 Resolving Conflicting Situations

Identifying the various types of conflicts in research is just a first step in dealing with them. The act of resolving those conflicts come next, and will require different strategies, somewhat dependent on the type of conflict involved (see Resnik 2014, for further discussion).

On the one hand, competing interests (CI) and conflicts of ownership (COO) often demand strategies that rely more heavily on judicial solutions. The resolution of conflicts of interests (COI) and conflicts of values (COV), on the other hand, are sought in ways that rely on regulation, instruction and even mediation.

8.6.2   Disclosure of Financial or Other Conflicts of Interest

To the parties involved in research, including institutions, government agencies, journals, and research subjects, *disclosure* is paramount. Disclosure makes explicit and transparent the important details related to the interpretation, credibility, and value of the information presented. It can be used as a simple tool to counteract bias and restore trust.

Most journals require that authors report potential conflicts of interest, both during the online submission process as well as explicitly within the article. They further run a 'declaration of interest' at the end of their articles.

Some researchers argue that such disclosure policies are inadequate, however. For example, with regard to the development of the DSM 5 (Diagnostic and Statistical Manual of Mental Disorders, Fifth Edition), Cosgrove and Wheeler (2013) maintain that current approaches, particularly with regard to the transparency of conflicts of interests, are insufficient to solve the problem of industry's grip on organized psychiatry, which is simply too powerful to be countered by 'disclosure measure(s).' They argue a more committal policy be put in place.

8.6.3   Regulation and Management of Competing Interests

In order to regulate and manage these issues, special committees or independent bodies have been founded that regulate and oversee projects involving conflicts of interest, flexing authority to approve or deny applications. Institutional Review Boards (IRBs) can fulfil this function, as they oftentimes do, and they too will demand the disclosure of any potential COIs that researchers may have. Agencies that dole out research grants may require that applications meet certain non-COI criteria, specifying, for example, source or amount of payment or service that is admissible. Finally, the installment of complaint offices, or trust committees may have a further preventative effect.

8.6.4   Prohibition and Penalizing Researchers Who Violate Disclosure Policies

Some argue that the role of institutions as the primary means for managing conflicts of interest should be reduced. This is because institutions have become players in the field and run the risk of harboring conflicts of interest themselves. Instead, as

Resnik (2007) argues, strict rules should be imposed that limits potential COIs, like research funding or stock ownership. Further, these rules must carry penalties for those who break them.

On the other hand, a complete prohibition of any possible conflict of interest could deprive science and society of important benefits. Indeed, [forbidding] universities from taking funds from industry would stifle collaboration between academia and industry' (Resnik 2007, p. 124).

8.6.5 Resolution in Court

Sometimes conflicts can only be settled in court. Regrettable as that may seem, it can also shed light on certain questions, because ownership rights come in the form of copyrights and patents, and they hold legal weight. In cases when the owners (sponsors) of research determine what may or may not be disseminated, challenging their decisions in court may be a last resort to settle differences of opinion, though they will be expensive for individual researchers or research institutes. However, it can be expected that these types of conflicts will increase as academics collaborate more with thirds parties and challenges over the rights to the knowledge at hand ramps up concurrently.

8.6.6 What to Make of All This?

Conflicting interests cast doubt on researchers motives and they have an undermining effect on overall trust in science. *However, not all conflicting interests result in COIs.* The question is *how* special interests impact a scientists' research, and more specifically, *how open* scientists are about that impact. By disclosing a researcher's affiliations, external positions, and (financial) ties, conflicts of interest and value can be avoided. That being said, additional regulation and legislation will also be necessary in the future.

8.7 Conclusions

8.7.1 Summary

In this chapter, we've looked at science from the perspective of competing markets. We highlighted how different parties (in particular, industry, and government) and their differing (commercial) interests hold influence over research agendas. We examined the valorization of scientific knowledge (focus on economic value of science) and the emergence of academic capitalism in the past 30 years.

We differentiated between conflicts of interest (COI) and conflicts of values (COV), and discussed their potential to cloud a researcher's judgement, but we also noted that not all interests lead to conflict situations. Disclosure of a researcher's affiliations, external positions, and (financial) ties, as well as further regulation may help resolve future COIs.

In addition, we proposed to use the term competing interests (CI) when looking into situations where researchers have to strike a balance between the interests of collaborators. Finally, we explored considerations within conflicts of ownership (COO), where researchers are not at liberty to report on their research as they believe fit, because they don't own it themselves. Copyright and patenting restrictions are at the base of these conflicts.

8.7.2 Discussion

While we investigated the reality of conflicts of interest in research, we neglected to examine the forces that drive the functioning of science itself. Under what conditions do researchers actually do their work? How are they affected by political considerations, or by administrative decisions, in practice? These questions will be highlighted in the next chapter.

Case Study: Complicit to Torture?

Near the end of World War Two, the US government felt the need to set up a program to help teach soldiers how to cope with torture, were they to find themselves captured. This program, known as SERE (Survival, Evasion, Resistance, and Escape), was extended through the Vietnam War era, and became relevant again after the 9/11 attacks and the ensuing wars in Iraq and Afghanistan.

Decades after implementation, it was realized that reverse-engineering SERE techniques could be used to interrogate detainees. Psychiatrist Paul Burney and psychologist John Leso (with assistance from an unnamed technician) were instructed to form a Behavioral Science Consultation Team (BSCT) and ramp up intelligence collection. In June of 2002, they were sent (under highly contested conditions) to Guantánamo Bay, the US detainment camp on the shores of Cuba that held over 700 prisoners suspected of terrorism.

As explored by Sheri Fink in her 2009 exposé 'Tortured Profession' (from which we draw the following details, including the quotations from several of the people involved), Burney and Leso seemed to have transgressed ethical boundaries when they devised this reverse-engineered interrogation method for the US Army. Burney and Leso had no formal training in interrogation and there were no standard operating procedures in place to guide their work. They confessed that they did not know what they were supposed to do, and therefore contacted psychologist Larry James for guidance. James had served as an Army psychiatrist at Abu Ghraib (the infamous prison in Iraq where gruesome violations of human rights by US military personnel later took place against detainees). Apparently, James too felt unsure of how to proceed, and turned to Louis Banks for guidance. Banks was experienced with preparing soldiers to deal with mild 'forms of torture,' including slapping, pushing (into walls for example), forced placement into uncomfortable positions, and waterboarding (simulating the feeling of drowning).

Within just 5 months of their arrival at the Guantánamo facility, Burney and Leso prepared a memo. In it, they proposed several interrogation techniques using both physical and psychological pressure, including waterboarding, sleep deprivation, noise exposure, and limited access to comfort objects such as the Quran (Fig. 8.3).

They furthermore offered three approaches for different categories of prisoners, with the third reserved for 'high-priority detainees' who showed 'advanced resistance.' These prisoners could be subjected to food restriction for 24 h, once a week. They could also be led to believe they would be subjected to experiences with 'painful or fatal outcomes.' Furthermore, they could be exposed to cold weather or water 'until such time as the detainee began to shiver.'

Fig. 8.3 Francisco de *Goya: Why?* Etching, from the series 'The disasters of war' (1810–1820)

However, the authors warned in their memo that 'physical and/or emotional harm from the above techniques may emerge months or even years after their use,' and that 'the most effective interrogation strategy is based on building rapport.' In fact, they continued, 'interrogation techniques that rely on physical or adverse consequences are likely to garner inaccurate information and create an increased level of resistance.' They had fewer reservations regarding the use of psychological pressure, such as sleep deprivation, withholding food, and isolation, which they deemed 'extremely effective.' They further noted that it was 'vital' to disrupt normal camp operations through the creation of 'controlled chaos' (quoted in Fink 2009).

Banks (the experienced Army psychiatrist) who received a copy of the memo, expressed his own misgivings about the use of physical pressures, which he believed could bring about 'a large number of potential negative side effects.' For one, when 'individuals are gradually exposed to increasing levels of discomfort, it is more common for them to resist harder.' Further, 'it usually decreases the reliability of the information.'

Despite these concerns, the proposed techniques were put to use interrogating prisoners at the Guantánamo facility. Among those subjected to interrogation was Mohammed el-Khatani, a high-value detainee suspected of involvement in the 9/11 attacks. Psychologist Michael Gelles, working with the Guantánamo Criminal Investigation Task Force, criticized the interrogation methods, arguing that they weren't up to his professional standards. Even the FBI objected, believing them to be 'illegal.' These protests were all ignored (Fig. 8.4).

Fig. 8.4 Detainee at the Abu Ghraib prison, with bag over their head, standing on a box with wires attached, Nov. 4, 2003. (Source: public domain)

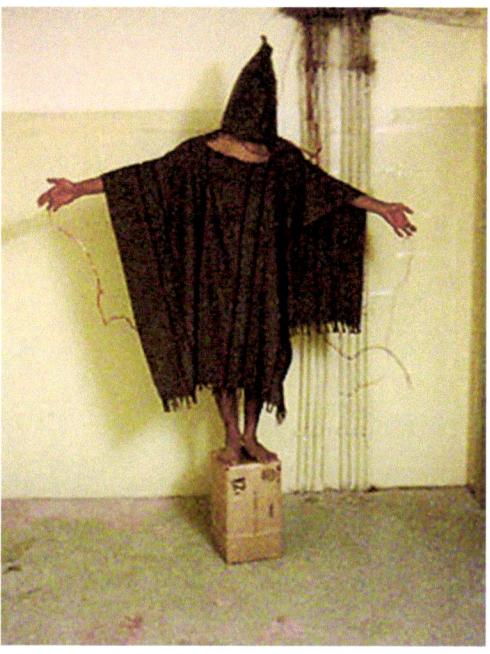

In June 2005, a log of el-Khatani's interrogation was published by *Time Magazine*, causing a good deal of public indignation (Zagrin 2005). Bioethicists condemned the procedures as a clear violation of the 'golden rule' in bioethics (do not harm, contribute to beneficence). A week later, the American Psychological Association (APA) set up a ten-member committee (with six members having worked or consulted for the military or the CIA) to investigate the ethical aspects of the case.

The APA ruled that it was 'consistent with the APA Ethics Code for psychologists to consult with interrogators in the interests of national security.' However, psychologists should themselves not participate in torture, and 'have a responsibility to report it.' Practicing psychologists are 'committed to the APA ethics code whenever they encounter conflicts between ethics and law,' and if a conflict cannot be resolved, 'psychologists may adhere to the requirements of the law.'

Banks and James (both members of the APA) also defended the Burney-Leso team (who drafted the memo), arguing that these psychologists had helped fix the problems rather than cause them. In 2005, James wrote: 'the fact of the matter is that since Jan 2003, wherever we have had psychologists no abuses have been reported' (quoted in Fink 2009).

The APA's lenient attitude ultimately backfired. Two years later, a petition by APA members was accepted that banned psychologists from working in detention settings where international law or the U.S. Constitution was violated.

Of course, since this case broke and caught public attention, many more details about the conditions at Guantánamo have come to light. Legal battles have been fought in court over the legality of detaining prisoners under such poor conditions, with little access to counsel and most of them having never been tried. Attempts by President Obama to close Guantánamo were unsuccessful. As of early 2020, some 40 detainees remain.

Assignment

Consider this case from the perspective of conflicts of interest, conflicts of values, and competing interests.

1. Identify any COI, COV and/or CI. Which values, norms, or interests are at stake in this case?
2. Consider whether you find Burney and Leso liable of professional misconduct. Are they responsible for facilitating unethical (irresponsible) behavior towards detainees, or should their expression of doubt about the use of such harsh interrogation procedures be regarded as a protection against liability?
3. What recommendations would you give to a psychologist who is asked to engage in a task they feel to be unethical?

Suggested Reading

For a critical appraisal of the corrupting effects of financial conflicts of interest on scientific norms, we recommend Derek Bok, *Universities in the Marketplace* (2003), Jennifer Washburn, *University, Inc.* (2005), David Resnik, *The Price of Truth* (2007), and Daniel Greenberg, *Science for Sale* (2009), though these sources mostly target the pharmaceutical and biomedical sciences. A similar study for the social sciences is a need we hope a future scholar will address. On the subject of 'valorizing science' we direct readers to the very informative, and short piece by De Jonge and Louwaars (2009). For an illuminating perspective on conflicts of interest, we refer to Resnik (1998).

References

Abdoul, H., Perrey, C., Tubach, F., Amiel, P., Durand-Zaleski, I., & Alberti, C. (2012). Non-financial conflicts of interest in academic grant evaluation: A qualitative study of multiple stakeholders in France. *PLoS One, 7*(4), e35247. https://doi.org/10.1371/journal.pone.0035247.

Angell, M. (2004). *The truth about drug companies: How they deceive us and what to do about it.* New York: Random House.

Aubert, V. (1963). Competition and dissensus: Two types of conflict and of conflict resolution. *Journal of Conflict Resolution, 7*(1), 26–42. https://doi.org/10.1177/002200276300700105.

Bartelson, J. (2000). Three concepts of globalization. *International Sociology, 15*(2), 180–196.

Bero, L. A., & Grundy, Q. (2016). Why having a (Nonfinancial) interest is not a conflict of interest. *PLoS Biology, 14*(12), e2001221. https://doi.org/10.1371/journal.pbio.2001221.

Bok, D. (2003). *Universities in the marketplace. The commercialization of higher education.* Princeton/Oxford: Princeton Universities Press.

Chivers, T. (2019, July 4). Does psychology have a conflict-of-interest problem? *Nature, 571,* 21–23.

Cosgrove, L., & Wheeler, E. E. (2013). Industry's colonization of psychiatry: Ethical and practical implications of financial conflicts of interest in the DSM-5. *Feminism & Psychology, 23*(1), 93–106. https://doi.org/10.1177/0959353512467972.

Cosgrove, L., Krimsky, S., Vijayaraghavan, M., & Schneider, L. (2006). Financial ties between DSM-IV panel members and the pharmaceutical industry. *Psychotherapy and Psychosomatics, 75*(3), 154–160. https://doi.org/10.1159/000091772.

Davis, M. (2001). Introduction. In M. Davis & A. Stark (Eds.), *Conflict of interest in the professions* (pp. 3–19). Oxford: Oxford University Press.

De Jonge, B., & Louwaars, N. (2009). Valorizing science: Whose values? *EMBO Reports, 10*(6), 535–539. https://doi.org/10.1038/embor.2009.113.

Fine, B., & Saad-Filho, A. (2017). Thirteen things you need to know about neoliberalism. *Critical Sociology, 43*(4–5), 685–706. https://doi.org/10.1177/0896920516655387.

Goldacre, B. (2013). *Bad pharma: How drug companies mislead doctors and harm patients.* London: Faber & Faber.

Greenberg, D. S. (2009). *Science for sale: The perils, rewards and delusions of campus capitalism.* Chicago: University of Chicago Press.

Head, B. W., & Alford, J. (2015). Wicked problems: Implications for public policy and management. *Administration & Society, 47*(6), 711–739. https://doi.org/10.1177/0095399713481601.

Healy, D. (2002). *The creation of psychopharmacology.* Cambridge, Mass: Harvard University Press.

Hermanowicz, J. C. (2016). Universities, academic careers, and the valorization of "shiny things". In *The university under pressure (research in the sociology of organizations, Vol. 46)* (pp. 303–328). Emerald Group Publishing Limited.

Hoekman, J. (2012). *Science in an age of globalisation: The geography of research collaboration and its effect on scientific publishing (diss) Eindhoven*. Technische Universiteit Eindhoven.

Hari, J. (2018, Jan 7). Is everything you think you know about depression wrong? The Observer.

Krimsky, S. (2012). Do financial conflicts of interest bias research? An inquiry into the 'funding effect' hypothesis. *Science, Technology, & Human Values, 38*(4), 566–587. https://doi.org/10.1177/0162243912456271.

MacCoun, R. J. (2015). The epistemic contract: Fostering an appropriate level of public trust in experts. In B. H. Bornstein & A. J. Tomkins (Eds.), *Motivating cooperation and compliance with authority. The role of institutional trust* (pp. 191–214). London: Springer. https://doi.org/10.1007/978-3-319-16151-8_9.

Maj, M. (2008). Non-financial conflicts of interests in psychiatric research and practice. *British Journal of Psychiatry, 193*(2), 91–92. https://doi.org/10.1192/bjp.bp.108.049361.

Martens, M. P., Herman, K. C., Takamatsu, S. K., Schmidt, L. R., Herring, T. E., Labuschagne, Z., & McAfee, N. W. (2016). An update on the status of sponsored research in counseling psychology. *The Counseling Psychologist, 44*(4), 450–478. https://doi.org/10.1177/0011000015626271.

Mauss, M. (1970). *The gift. Forms and functions of exchange in archaic societies*. London: Cohen & West Ltd.

Mosbah-Natanson, S., & Gingras, Y. (2013). The globalization of social sciences? Evidence from a quantitative analysis of 30 years of production, collaboration and citations in the social sciences (1980–2009). *Current Sociology, 62*(5), 626–646. https://doi.org/10.1177/0011392113498866.

Resnik, D. B. (1998). *The ethics of science. An introduction*. London: Routledge.

Resnik, D. B. (2007). *The price of truth. How money affects the norms of science*. Oxford: Oxford University Press.

Resnik, B. (2014). Science and money: Problems and solutions. *Journal of Microbiology and Biology Education, 15*(2), 159–161. https://doi.org/10.1128/jmbe.v15i2.792.

Saunders, D. (2007). The impact of neoliberalism on college students. *Journal of College and Character, 8*(5), 1–9. https://doi.org/10.2202/1940-1639.1620.

Sherry, J. F. (1983). Gift giving in anthropological perspective. *Journal of Consumer Research, 10*(2), 157–168. https://doi.org/10.1086/208956.

Slaughter, S., & Rhoades, G. (2004). *Academic capitalism and the new economy: Markets, state and higher education*. Baltimore: Johns Hopkins University Press.

Song, F., Parekh, S., Hooper, L., Loke, Y. K., Ryder, J., Sutton, A. J., Hing, C., Kwok, C. S., Pang, C., & Harvey, I. (2010). Dissemination and publication of research findings: An updated review of related biases. *Health Technology Assessment, 14*(8), 1–193. https://doi.org/10.3310/hta14080.

Soudijn, K. (2012). *Ethische codes voor psychologen [Ethical code for psychologists]* (3rd ed.). Amsterdam: Nieuwezijds BV.

Turner, C., & Spilich, G. J. (1997). Research into smoking or nicotine and human cognitive performances: Does the source of funding make a difference? *Addiction, 2*(11), 1423–1426. https://doi.org/10.1111/j.1360-0443.1997.tb02863.x.

Volochinsky, D. P., Soto, R. V., & Winkler, M. I. (2018). Gifts and conflict of interest: In shades of gray. *Acta Bioethica, 24*(1), 95–104. Retrieved from https://revistaidiem.uchile.cl/index.php/AB/article/view/49382/57559.

Washburn, J. (2005). *University inc. the corporate corruption of higher education*. Perseus Books: New York.

Wiersma, M., Kerridge, I., & Lipworth, W. (2017). The dangers of neglecting non-financial conflicts of interest in health and medicine. *Journal of Medical Ethics, 44*(5), 319–322. https://doi.org/10.1136/medethics-2017-104530.

Yates, B. T., & Taub, J. (2003). Assessing the costs, benefits, cost-effectiveness, and cost-benefit of psychological assessment: We should, we can, and Here's how. *Psychological Assessment, 15*(4), 478–495. https://doi.org/10.1037/1040-3590.15.4.478.
Zhuang, G., & Tsang, A. S. (2008). A study on ethically problematic selling methods in China with a broaden concept of gray-marketing. *Journal of Business Ethics, 79*(1–2), 85–101. https://doi.org/10.1007/s10551-007-9397-1.

References for Case Study: Complicit to Torture?

Fink, S. (2009, May 5). *Tortured profession: Psychologists warned of abusive interrogation, then helped craft them.* https://www.propublica.org/article/tortured-profession-psychologists-warned-of-abusive-interrogations-505
Zagrin, A. (2005, June 20) Inside the interrogation of detainee 063. *Time Magazine.*

Chapter 9
Science and University Politics

Contents

Electronic Supplementary Material: The online version of this chapter (https://doi.org/10.1007/978-3-030-48415-6_9) contains supplementary material, which is available to authorized users.

© The Author(s) 2020

J. Bos, *Research Ethics for Students in the Social Sciences*,
https://doi.org/10.1007/978-3-030-48415-6_9

After Reading This Chapter, You Will:

- Understand how political factors impact modern science
- Appreciate in what ways the replication crisis endangers the values of science
- Know how publication pressure and perverse incentives challenge scientific practices
- See why teaching ethics requires reactive, proactive, and reflexive education

Keywords Adjunctification of higher education · Competition for prestige · Degree of freedom · False positives · h-index · Impact factor · Matthew-effect · Motivation (intrinsic v/s extrinsic) · New public management · Nomothetic · Paradigm · Performance indicators · Performance-based research funding system (prfs) · Perverse incentives · Publication bias · Publication pressure · Research rigor · Replication (exact v/s conceptual) · Replication crisis · Salami slicing technique · System theory

9.1 Introduction

9.1.1 Is There a Crisis in the Social Sciences?

The Guardian, a leading British newspaper, had a long running series on abuses and exploitation in academia titled 'Academic anonymous'. From this series, we draw a few examples. In one piece (30 June 2017), the work of an unnamed researcher is described. It shows how the researcher tailored their papers not to fit the data, but to make the papers tailor fit for publication in prestigious journals. Another researcher left out any data interpretations that could raise questions with the journal editors. Their supervisor told them it could lead reviewers to turn it down. In yet another example, a research supervisor informed their students that research begins not with asking questions but with selecting suitable, high-ranking journals and defining subjects that might fit into those journals. All these examples illustrate the less than perfect publication behavior currently practiced in some corners of academia.

It was precisely experience with such practices that incited cognitive neuroscientist Chris Chambers to write *The Seven Deadly Sins of Psychology* (2017), an indictment of the numerous research misbehaviors found in the field of psychology, many of which already have been discussed in previous chapters of this book (see especially Chaps. 6 and 7), ranging from forms of bias and unreliability to fraud or even corruption.

After working for nearly 10 years in the field, Chambers in an op-ed for *The Guardian* (May 9th 2017), said that he understood that a psychologist's mission wasn't about truth seeking, it was about 'crunching through as many experiments as possible as quickly as possible, finding ways to make ambiguous data look beautiful, publishing frequently in prestigious journals [...] winning large public grants, and basically getting as famous and powerful as possible'.

Whether we will call the state of affair psychology finds itself in a 'crisis,' as some do, or more of a 'challenge,' may be a matter of opinion. Chambers found that not enough had been done to address the forces that fueled the problem (see our discussion of 'academic capitalism' in Chap. 8). An appeal to research ethics could be instrumental in transforming the field psychology in helping it to overcome this (perceived) crisis.

Whether this is the case will be the subject of debate in this chapter. We realize, however, that many of the discussions in this chapter are not part of a student's lived experience. Many of the themes addressed here may seem 'abstract,' far removed from the reality of everyday life. To a degree this is true, but to a degree it isn't, because the pressures and incentives, and political considerations that make up the fabric of modern academic life will reveal themselves in the educational practices in which students are immersed. This is why we have chosen to dedicate one chapter of our book to some of these broader topics.

9.1.2 System Approach

In this chapter, we will broaden our perspective on research ethics, and move away from considering how or why individual researchers decide to break from established norms and values, or from examining what happens when rigid guidelines meant to steer our ethical behaviors are not followed. Instead, we will look at the interrelated and interdependent parts that make up the whole of the scientific enterprise, of which ethics is just one element. This holistic approach is often referred to as a *systems perspective* (see Parson 1951; Wiener 1965; Maturena and Varela 1980). By zooming out to a lens that can capture the system as a whole, we can consider how scientific practices and ethics mutually influence one another. This allows us to explore some of the most important internal factors that regulate scientific operations and shape our understanding of how ethics play a role.

We begin with an exploration of what is currently considered one of the more challenging problems in the social sciences, namely the *replication crisis*. Next, we will research two factors that are often closely connected to said crisis, both believed to have a corroding influence on ethics: *publication pressure* and *perverse incentives*. In the final section, we will focus on the role of teaching research ethics in the future as a means of countering the corroding effects of misconduct.

9.2   Replication Crisis

9.2.1   What Is a 'Replication Crisis'?

The term *replication crisis* came into circulation around 2010, when it was observed that often, when studies were reproduced, the same findings could not be replicated. The replicated studies would find a smaller or larger effect than originally claimed, or none at all, or the direction of the effect had changed, or perhaps an entirely different effect was found altogether.

Derived from the natural sciences, the requirement that credible knowledge must consist of reproducible findings implies that studies carried out under *the same conditions should generate the same results*. Findings that cannot be reproduced are considered *chance findings, or false positives*, and do not belong in the scientific body of knowledge (Makel 2017).

The requirement of replicability is based on a *nomethetic* approach, meaning that it should be possible to discover laws that explain objective phenomena. Although not every social scientist accepts the requirement of replicability and its underlying nomothetic notion, those who do are committed to experimentation, clinical trials, and hypothesis testing under controlled conditions. The observation of a replication crisis calls into question the credibility of such research findings.

In fact, many of Daniel Kahneman's well known social priming studies (as detailed in *Thinking, Fast and Slow*, 2011) failed in replication, likely because the sample sizes had been too small. 'I placed too much faith in underpowered studies,' Nobel prize laurate Kahneman famously admitted (*Retraction Watch*, February 20th 2017). Kahnemann is but one of many whose studies failed to be replicated (see Hughes 2018, for more examples).

What constitutes the 'replication crisis' is a digression from the ideal of reproducibility, resulting in the view that many research findings in certain disciplines (notably psychology, but also medicine) are unreliable, or perhaps even false. To better grasp the issues at hand, we discuss three aspects of the replication crisis: (1) lack of replication studies, (2) weak research rigor, and (3) failing replication.

9.2.2   Lack of Replication Studies

While replicability is accepted as a 'fundamental principle' by a large portion of the social science community, very few replication studies are actually carried out. This was noticed by Smith as early as 1970, while more recently, Makel et al. (2012) found an overall replication rate of psychological research at just over 1%. Additionally, many of these replication studies were performed by the original author. Why is there so little interest in replication studies? Let's look into two possible reasons.

One rationale is *theoretical sectarianism* (favoritism of one's own theories, and prejudice or discrimination against rival ones), to which many social scientific disciplines are prone. Theoretical sectarianism occurs because researchers operate under specific theoretical assumptions, or *paradigms,* as they were defined by Thomas Kuhn (1970). Kuhn conceptualized a change in theoretical assumptions as a *paradigm shift.* The problem being, paradigm shifts aren't often universal, with different disciplines regularly operating under different paradigms. Furthermore, there is often little common ground to be found between differing paradigms.

Without common ground (shared theories, mutually referenced authors, agreed upon methodologies), researchers are more inclined to confirm findings in their own domain than to disconfirm those of their peers in other domains. This implies that the use of unfamiliar methods make an interdisciplinary researcher performing a replication study to be more vulnerable to the possibility (or accusation) of misunderstanding (see Box 9.1 for an example thereof).

Box 9.1: Irreproducible Tears?

The issue of 14 January 2011 of the journal *Science* ran a paper by a team of Israeli scholars from the Weizmann Institute of Science at Rehovot, that claimed that human tears my serve a chemosignaling function. After a team of Dutch scholars subsequently failed in their attempt to replicate the experiment described in the original article, a dispute broke out between the two groups of scientists, each accusing the other of misrepresentation. The case, which we describe below in some detail, illustrates the difficulties of replication in the social sciences.

The Israeli team, headed by Naom Sobel, hypothesized that human tears serve a *chemosignaling function*, such that men smelling or breathing in female tears become sexually less aroused. The idea is that tears contain certain chemicals that may influence brain activity in (heterosexual) men. In order to test this hypothesis, the researchers collected 'donor tears' from women who had watched sad films. They then exposed a group of males in a *within-subjects design* to these tears as well as to a substitute substance. After exposure, the participants were asked to rate their sexual attraction to pictures of female faces. In a second experiment, the researchers added another dependent variable, namely levels of psychophysiological arousal. The various studies revealed that exposure to female tears indeed has an effect, both on self-reported levels of sexual arousal (modest) and on objective psychophysiological expression (more pronounced) (Gelstein et al. 2011).

A group of mainly Dutch psychologists attempted to replicate these findings. One experiment was set up as an exact replication, the other two had

(continued)

Box 9.1 (continued)

alterations. Thus, in one condition, male participants were not only asked to rate sexual attraction, but also whether or not they would be willing to date the females in the pictures. In a second series of experiments, the researchers changed the design from a within-subjects study (which had low power), to a larger, *combined within and between subject design*. However, they now asked the subjects to not only rate the pictures of a female face, but also of the whole body as a measure of 'arousal.'

Their attempts to 'replicate and extend' the original studies failed. They found no support for the chemosignaling function hypothesis, which they considered a possible *false finding* (Gračanin et al. 2017a, b, p. 149). The authors proposed instead that tears are functional in a *social* context, and that crying is a 'self-soothing strategy' (Gračanin et al. 2014).

Principal investigator Naom Sobel (2017) of the original Israeli study responded to the failed attempt at replication, arguing that the replication studies were not really replications of the original at all. The researchers had not operated from a proper chemosignaling laboratory, he argued. Further, he stated they had used different test materials (other films), which communicated a different feeling (not sadness), and had used combined datasets in an 'inappropriate manner.' Had they used the 'appropriate techniques,' Sobel believed they would have found that the data the Dutch team collected actually supported the original hypothesis. This prompted a further response from the Dutch replication team (Gračanin et al. 2017a, b), who argued that a theory that only holds under very specific circumstances is likely not a very good theory.

This dispute is of interest because it leaves open any of the following alternatives, namely that this 'failure to replicate' represents:

1. An instance of theoretical sectarianism
2. The social sciences' context dependency
3. Weak research rigor

Which of these alternatives do you believe is most likely?

A second rationale may be that journals prefer to publish 'newsworthy findings' over what is 'already known' (see the section on dissemination bias in Chap. 6 for further discussion). For this same reason, it is more difficult to get funding for replication studies, especially *exact replication studies* (same methodology, same conditions), as opposed to *conceptual replication studies* (same conceptualization, different methodology or different conditions).

The net result is that in the absence of adequate replication studies, we are much less sure that positive findings are in fact positive, and not accidental (see Fanelli 2010a for a discussion).

9.2.3 Weak Research Rigor

The replication crisis may be further worsened by *weak research rigor*. In 2011, Simmons et al. published a now famous study that shows how easy it is to produce false positives if the researcher's *degree of freedom* is manipulated. Degrees of freedom pertain to a series of methodological choices researchers must make over the course of collecting and analyzing data, including the selection of dependent variables, determining sample size, using covariates (independent variables), and reporting subsets of experimental conditions (Simmons et al. 2011).

While staying within the accepted boundary of a .05 p-value, the authors succeeded in 'proving,' in a real experimental study with real subjects, the unlikely conclusion that listening to certain types of music makes you feel younger, and the obviously false conclusion that listening to certain other types of music *actually* makes you younger. They arrived at these conclusions by tacitly manipulating the researcher's degrees of freedom (such as changing the number of participants without reporting this, or reporting on certain measures only). A reviewer unaware of these manipulations would have to accept the 'age effect' (that you actually become younger by listening to certain music) as genuine.

The point the authors sought to make was that as long as reviewers and readers are not informed about the researchers' choices within their degrees of freedom (which can in and by themselves be legitimate), they cannot reasonably separate false from true findings. They therefore recommend more transparency about the choices made by researchers to avoid the creation of false positives.

9.2.4 Failing Replication?

Another seemingly significant blow was delivered in 2015, when a number of researchers undertook a large scale attempt to replicate a swath of psychological studies. The result were sobering. At the insistence of Brian Nosek, a group of 270 authors (worldwide) replicated one hundred psychological studies, published in the course of 1 year (2008), in three top ranking journals (Open Science Collaboration 2015).

Using a uniform replication protocol, all studies were carried out as *exact replications* as much as possible. The results were collected, assessed, and independently reviewed. Comparing the original studies with the replication studies, the researchers looked at significance and p-values, effect size, subjective assessments, and meta-analyses of effect size. What they found was that collectively, these indicators revealed that the replications produced significantly weaker evidence than the original findings. For example, the mean size effect of the replication studies were roughly half the magnitude of original the mean size effect (see Fig. 9.1).

The Open Science Collaboration project caused shockwaves that resonated beyond the scientific community. Newspapers all over the world reported that psychological studies fail to replicate. The authors themselves were much more careful,

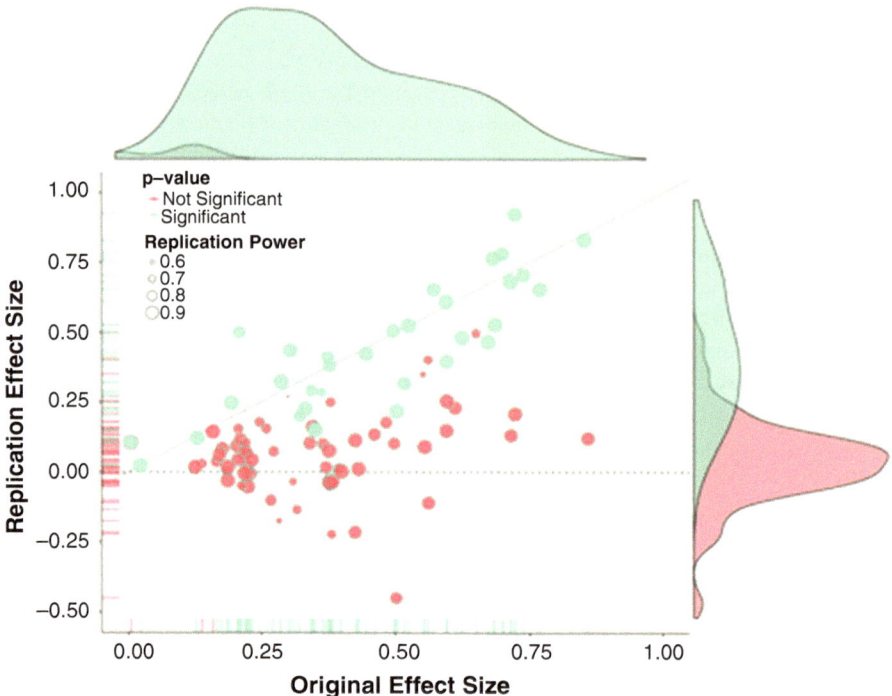

Fig. 9.1 Original study effect size versus replication effect size. (© *Science*, 28 Aug. 2015, issue 6251)

however. They argued that their study neither proved nor disproved anything. 'The original studies examined here offered tentative evidence; the replications we conducted offered additional, confirmatory evidence. In some cases, the replications increase confidence in the reliability of the original results; in other cases, the replications suggest that more investigation is needed to establish the validity of the original findings. Scientific progress is a cumulative process of uncertainty reduction that can only succeed if science itself remains the greatest skeptic of its explanatory claims' (Open Science Collaboration 2015, p. 4716–7).

Maxwell et al. (2015) similarly argued that when replication studies fail to show significant results, this should not lead to the premature conclusion that the original study was somehow faulty or flawed. In a reanalysis of the data collected by the Open Science team, Van Bavel et al. (2016) were able to demonstrate that some of the failure to reproduce the same findings could be attributed to *contextual factors*, having to do with social and cultural differences between the countries where the studies were carried out, illustrating just how difficult it is to replicate social science studies.

9.3 Publication Pressure

9.3.1 Value for Money

In Chap. 8, we discussed how in the last quarter of the twentieth century, globalization and neoliberalism set in motion a trend towards 'valorization' of science, that emphasizes the monetary value of scientific knowledge. Ties with external parties (both commercial organizations and governmental bodies) strengthened, from which a variety of conflicts emerged. But 'valorization' also impacted the way science was organized internally, including publication behavior, to which we now turn our attention.

The wish to demonstrate to the tax paying public that scientists produce 'value for money' found its way into an administrative logic that demands adherence to uniform, quantifiable norms. University administrators and policy makers began thinking about ways to measure a researcher's output in terms of 'objective numbers'. One objective criterion they soon came up with was a count of articles published by individual researchers in prestigious journals. Later this was extended to a count of citations received. Both could be used as an approximation of a researcher's objective impact on the scientific community.

These measures resulted in the establishment of new standards, such as a journal's *impact factor*, a researcher's *citation impact*, and their so-called *h-index* (see Box 9.2). Any of these standards, with a special focus on publishing success, became a measuring stick for success in academia (Barnard-Brak et al. 2011). Greater output and more citations equaled greater academic value. Unwittingly, these norms transformed the tenured job market in academia into a 'market for prestige' (Garvin

---

Box 9.2: H-index (Adapted from Wikipedia)
The *h-index* (named after its inventor, Jorge Hirsch) is an author-level metric that attempts to measure both the productivity and citation impact of a scientist's publications. The index is based on the scientist's most cited papers and the number of times those papers were cited in other publications. The index can also be applied to the productivity and impact of a scholarly journal, as well as a group of scientists residing within the same department, university, or even country. The values on the h-index vary greatly from discipline to discipline, where the number of scientists, published papers, and citations strongly differ. In physics, for example, the average h-value for a fulltime professor is between 15 and 20, in economics its 7.6, in sociology its 3.7. In many academic communities, it was long believed that better scholarship equaled higher h-factors, but recent discussion flared up about whether h-values actually represent quality.

Fig. 9.2 Publication pressure

1980), thereby creating of whole series of new questions, which we will outline below (Fig. 9.2).

9.3.2   *Publish or Perish*

Has the pressure to publish in 'high impact journals' altered research norms? Moustafa (2014, p. 139) observed that the impact factor became 'a major detrimental factor of quality, creating huge pressures on authors, editors, stakeholders and funders.' More tragically, Moustafa notes, is that impact factor has also become the condition for allocating government funding to entire institutions in some countries.

Siegel and Baveye (2010) reason that scholars who wish to meet publication expectations will resort to a variety of techniques to increase their output and crank up their citation ranking. These include the use of *co-authorships* and so called *gift authorships* (author does not contribute to the research, or not significantly, but is included out of courtesy and in the expectation of reciprocity). Additionally, *salami slicing* techniques (slicing research such that several different papers can be written, all slightly varying around the same theme) and *extensive referencing* ('I cite-you, you-cite-me') are employed to meet publication expectations (Box 9.3).

Fanelli and Larivière (2016) researched why publication rates have increased in the last quarter of the twentieth century. Looking into the work of over 40,000 researches in Western countries, whose profiles were drawn from the Web of Science, they compared the publishing frequency between 1900 and 1998. They found that the average number of papers published throughout the twentieth century remained stable for most disciplines, and then visibly increased after 1980. However,

Box 9.3: Slicing and Dicing: A Dilemma

A well-respected colleague proudly explains to you that he has managed to produce 12 publications out of the one dataset he collected for his dissertation. This is a particularly interesting achievement, as it involves a dataset with only 232 respondents to a four-page survey.

How do you respond?

1. I think this is a great example to follow and I ask him how he achieved it.
2. I cannot imagine each of these 12 papers having a unique contribution and vow never to go down this route.
3. I tell the colleague that this is bad science and that I strongly disapprove of their actions.
4. I think this is bad practice and is tainting the reputation of science. I inform the editors of at least the most recent of the 12 publications.

(Adapted from the *Erasmus Dilemma Game*)

so did the number of co-authors. *Fractional productivity* (the productivity of one researcher's work spread out over multiple co-authored papers) remained stable. From this, the authors concluded that the widespread belief that pressure to publish causes the scientific literature to be flooded with salami-sliced, trivial, and false results is in fact incorrect, or at least exaggerated.

In a similar examination, van Wesel (2016) looked at output of journals rather than individual researchers. He examined 50 high impact scientific publication outlets selected from a variety of disciplines (including medicine, physiology, and psychology), to see whether publication behavior changed between 1997 and 2012. Van Wesel compared the number of authors listed, the amount of references included, but also the text and abstract lengths, and even the presence of a colon in the title (as indicators of a paper's 'citability'). He found similar patterns as Fanelli and Larivière (2016), including a growing number of co-authors and increased referencing. Aware that these patterns could also be attributed to a change in editorial policies, van Wesel (2016, p. 212-3) believes it is nevertheless 'not unrealistic to link the observed changes in publication behavior to a change in evaluation criteria.'

While the studies discussed above do not provide evidence that 'pressure to publish' leads to fraudulent behavior (Fanelli 2010b), there is reason to believe that publication norms have changed. Authors seem much more strategic in their publication behavior. Due to an awareness of the necessity to publish in order to further their academic careers, it seems likely that researchers plan their publications to meet these goals.

9.3.3   Intrinsic Versus Extrinsic Motivation

Does increasingly 'strategic' and 'planned' publishing behavior means that authors are less intrinsically motivated? Hangel and Schmidt-Pfister (2017) interviewed over 90 researchers in Germany who were in different stages of their careers (PhD students, post-doc, tenured professors). The authors asked their participants frankly: 'Why do you publish?' They found that the most common reason to publish contained both internal and external motivations, namely: (1) to communicate interesting research results; (2) to gain recognition among peers; (3) because they enjoy writing, and (4) to obtain funding in order to secure future research.

All researchers draw from a mix of these motivations, albeit differently in different stages of their careers. Those still in the early stages of their careers, when they are more dependent on their supervisors, are often highly aware of the incentive that one *has to* publish for mere survival. They cite motives 2 and 4 as most important. In the next stage of their career, many still feel the pressure to publish, but now more as a means to an end. These researchers seem more capable of enjoying the research process itself and more often cite reasons 1 and 3. Tenured professors almost always cite reason 4, claiming to publish for educational purposes and for academic survival.

9.4   Perverse Incentives

9.4.1   Reward Systems

Publication pressure represents one of the factors that contribute to what some believe to be a precarious situation in the social sciences. Ambition, external pressure, and weak research rigor seem to highlight an element of 'crisis' mentioned earlier in this chapter. Outside of publication pressure, the other factor consists of reward systems that aim to stimulate academic quality and simultaneously deal with decreased public funding.

Acting on a desire to transform universities along neoliberal lines into efficient, productive, and outstanding institutions, *new public management* administrators (see Box 9.4) set up national reward systems to select the best researches. *Performance-based Research Funding Systems* (PRFS) have since been used to evaluate the quality of research proposals. They are based on the rationale that 'funding should flow to the institutions where performance is manifest' (Herbst 2007, p. 90).

These systems take into account any number of 'performance indicators,' such as the output of a research group, their citations received, their international ranking, and the judgements of their peers. For individual researchers, the number of successful grant applications is counted, along with their level of participation in international research associations, the number of keynote addresses they've made, board memberships, awards bestowed, and even their perceived societal impact (as expressed, for example, in their role as advisors for social organizations).

Box 9.4: New Public Management
New Public Management (NPM) is the late twentieth century approach to public service organizations that suggested they be run like businesses. It is based on principles of expanding managerial freedom, flexibility of organizational structures, shifting staff and job conditions, emphasis on output and decrease of input ('cost-effective management'), and increase of efficiency (see Christopher Hood, 'A Public Management for All Seasons?', 1991).

Perhaps unsurprisingly, researchers faced with the prospect of losing funding started to seek out ways to increase their performance accordingly. In an attempt to keep pace, or even outdo their colleagues, researchers began spending more time writing research proposals, but since each of these have a limited chance of being honored, the 'battle for efficiency' effectively resulted in an even greater waste of public resources.

PRFSs and other performance-based evaluation systems have unintended, so called *perverse effects*, when they encourage behavior that runs contrary to the original intentions (stimulation of excellence resulting in inefficient use of public funding) or invigorate undesirable behavior (competition for prestige, see Hicks 2012). Bouter (2015, p. 157) paints a grim picture of how certain incentives lead to a 'monoculture focused on citation scores, short term economic gains, and government-defined growth sectors,' with young talented researchers not being scouted. In the following sections, we discuss areas where such incentives can have perverse effects (see also Stone 2002).

9.4.2 Matthew Effect

Sociologist Robert Merton (1968) reasoned how symbolic, as well as material rewards in science will have an accumulative effect. If a researcher 'scores' on any one of the criteria mentioned above, their chances of getting funded improve, and this effect will accumulate over time (also in a negative sense: if you don't score, your chances diminish accordingly). The result is that 'eminent scientists get disproportionately great credit for their contributions to science while relatively unknown ones tend to get disproportionately little for their occasionally comparable contributions' (Merton 1988, p. 607). This is called the *Matthew Effect*, named after the Gospel of Matthew 25:29: 'For to every one who has will more be given, and he will have abundance; but from him who has not, even what he has will be taken away.'

There is some evidence that indicators of past performance correlate positively with chances of getting a grant application funded. This has been found in the fields of economics and the social sciences in the Netherlands (van den Besselaar and

Leydesdorff 2009; Bornmann et al. 2010), although the correlation is low and is not consistent across all disciplines.

More worryingly, since the Matthew Effect contains an element of self-fulfilling prophesy in it, there is a danger of misrecognition attached to it, especially with young academics. When academics are overly assessed in the early stages of their careers, precocious students will have a far better chance of surviving the competitive struggle over their late blooming peers, who may be just as brilliant.

9.4.3 Gap Between Tenured and Contingent Faculty Members

A further concern is that certain policies will negatively affect the academic labor market (Schwartz 2014). There are strong indications that neoliberal university politics, though meant to reward academic excellence, contribute to a divide between tenured staff with permanent positions and contingent faculty members who work on temporary contracts. This process is referred to as the *adjunctification of higher education*, or academia's overreliance on temporary, non-tenured faculty members (Curnalia and Mermer 2018).

There are a number of reasons for this *adjunctification*. Increased focus on retention, career outcomes, and resource acquisition brought about a reduction in tenured positions. Once regarded as essential to protect academic freedom in the pursuit of knowledge, today tenured positions making up sometimes as little as 20 to 25% of all faculty staff in Western universities. The majority of instructors and teachers are hired on a contingent basis. These have little prospect of getting tenured but are assessed by and large along the same criteria as tenured staff, while their possibilities for doing research (and getting promoted) are diminishing. Teaching and research are thus increasingly being undertaken by different kinds of faculty (Finkelstein 2014). If nothing changes in the near future, it is to be expected that this divide will only grow deeper (see Dobbie and Robinson 2008) (Fig. 9.3).

In a survey among some 1500 higher education professors, deans, governing board members, campus administrators, policymakers, and other stakeholders in the United States, Kezar et al. (2015) found general agreement that the present system is untenable. It threatens academic freedom and undervalues teaching through its disproportionate emphasis on research. Most respondents agreed on the necessity to restructure teaching positions. More full-time faculty, differentiation of responsibilities, and an overarching need to restore professionalism to the role of faculty were among the most pressing urges uttered by administrators and professors alike.

Below, we give the floor to two adjunct faculty who experienced this gap between tenured and contingent up close. One, a Dutch scholar in the social sciences, left academia after years of working on a temporary basis. In an email to her colleagues (quoted here with permission), she wrote: 'I experience a large disconnect between what you are paid to do as a temporary staffer (teaching) and what determines your career options (research). This disconnect means that many contingent faculty members are almost forced to put in a lot of additional hours (i.e., evenings,

Fig. 9.3 If nothing changes, the gap will grow

weekends, using holidays to write papers, taking up parental/care leave to decrease the teaching load and use that time to work). If you do not want to go along with this, your research output will stay 'behind' compared to the competition, which makes you less attractive and decreases your chances on permanent positions, promotions, grants, etc.'

Similarly, an adjunct faculty member in England was quoted as saying: 'I've watched brilliant friends be employed for two or three consecutive years with demanding teaching loads, travelling to cities hundreds of miles away with sharing childcare, only to be dropped for someone else with a more illustrious publication record' (*The Guardian*, July 14th 2017).

9.4.4 Gender Gap?

Historically, women have been underrepresented in all disciplines in academia and at all stages in their careers. Many industrialized countries have adopted strong gender equality programs in research and innovation, and the gender gap has since grown smaller (Ceci et al. 2014). Despite this, there is still a 'pipeline leakage,' meaning that on the way to the top, women drop out more often than men (Huyer 2015). From the 2015 report *She Figures*, issued by the European Commission, we learn that in 2011 some 50% of all students in the social sciences and law in the EU, about 30–40% of researchers in these fields, and 29% 'top level' researchers therein are women (with considerable differences per country).

Is the demise of the gender gap in higher education (in industrialized countries) an effect of government policies or does it happen independently? Ginther and Kahn (2009) compared the careers of male and female researchers in the US between 1973 and 2001. They examined the probability of obtaining a tenure track after getting a PhD and found that women were still less likely to take tenure positions in science. This was explained by 'fertility decisions' rather than reward systems – meaning that 'women must face a choice between having children or succeeding in their scientific careers, while men do not face the same choices' (Ginther and Kahn 2009, p. 183).

In non-industrialized countries (for example in parts of Africa), the gender gap persists, and it may also not be ascribed to government policies but rather cultural expectations. Ogbogu (2011), reviewing the gender gap in higher education in Nigeria, notes that recruitment and selection practices in their universities do not discriminate against women. Instead, factors such as lack of mentoring, poor compensation, family responsibilities, 'and the ideology that women should have low career aspirations' accounted for the observed disparity in academia.

9.4.5 Summing Up

In an attempt to keep up with and respond to worldwide changes in the political and economic landscapes within which universities operate, university administrators and government policy makers have set up procedures that aim to enhance productivity and excellence. These policies have had (and continue to have) intended as well as unintended consequences, resulting in a number of fundamental changes in the ways universities conduct research and provide education. These changes themselves pose new questions and challenges that must be addressed (Fig. 9.4).

9.5   Teaching Research Ethics and Integrity

9.5.1   Ethics and Integrity Education

There is little doubt that ethics and integrity education is becoming increasingly important in universities, not least because of increasing demands and competition in the academic field (Brall et al. 2017). Universities have a commitment to prepare, guide, and mentor students through a litany of ethical issues; combatting scientific misconduct, addressing questionable research practices, applying specific procedures and regulations, learning to deal with newfound responsibilities belonging to certain roles, knowing how to accommodate a diversity of perspectives, and learning how to deal with external pressure (Naimi 2007). We concur with Resnik (1998, p. 174) that the question is not whether 'ethics be taught?', but 'how can ethics be taught?' In the sections below, we briefly discuss three broad approaches for how ethics can be taught.

Fig. 9.4 © Cartoon by
John Stuart Clark

9.5.2   Reactive Education

This approach is focused on the prevention of misconduct and misuse of procedures. To accomplish *reactive education*, many research institutions offer *case-
based approaches* in their curricula (with either real or hypothetical dilemmas),
where students learn to make judgement calls through structured discussions
(Sponholz 2000). These discussions allow students to 'evaluate conventions, define
responsibilities, articulate positions on different issues, and acquire some facility at
using a framework for ethical decision making' (Stern and Elliott 1997). Canary
(2007) shows that these approaches are successful in enhancing moral sensitivity,
moral judgements, moral motivation, and moral character in students.

One such development consists of the development of eLearning tools. In a discussion of the issues related to the emergence of computer-based learning, Esposito
(2012) finds that open networked learning environments 'encourage a participatory
research approach and therefore foster creative suggestions and shared solutions
from participants in an evolving landscape of ethical opportunities and challenges'
(p.323).

9.5.3   Proactive Education

This approach is focused on preparing students to actively participate in complex research environments and providing skills for adapting to changes in research policies. *Proactive education* makes use of role playing and simulation settings, which are used to train students to contribute to research themselves.

Sweet (1999) and Karkowski (2010) discuss the function of simulated review procedures with mock research proposals, used to prepare students to produce higher-quality research proposals themselves, with both authors finding these procedures helpful. Löfström (2016) discusses the use of role-playing strategies in academic integrity education, specifically staging panel discussions of realistic cases, which act as added value in facilitating perspective-taking and the broadening of a student's worldview.

9.5.4   Reflexive Education

This approach aims at developing ethical awareness not restricted to institutional procedures. Rather working within a broader definition of research ethics that includes social, political, and moral dimensions, Von Unger (2016) emphasizes the need for more critical dialogue in ethics education. As an example, a course format is discussed where sociology students were trained to reflect on a case that had political relevance. The students collected their own data, engaged in critical inquiry, learned to formulate and revise their own assumptions, and thus learned to become more self-critical, a cornerstone of ethical decision-making.

9.5.5   Should Misconduct Be Criminalized?

A final note on the question of what to do when teaching ethics and integrity fails to achieve its goals. Should misconduct be criminalized? Until very recently, misconduct rarely led to litigation. Even Diederick Stapel, whose case was discussed in Chap. 5, was never brought before a court, though he did lose his job and a large number of his articles were retracted.

A critical issue in deciding whether research misconduct should be subject to criminal law is how it is defined, argue Dal-Ré et al. (2020). Should it only cover well-known forms of fraud, such as plagiarism, fabrication, and falsifying, or should it extend to questionable research practices, such as selective reporting? This question is important, because while criminalization could deter everything that is regarded as research misconduct, it could simultaneously lead to normalization of what is *not* considered misconduct.

Dal-Ré et al. (2020, p. 9) admit that a research integrity organization with global authority will not emerge any time soon, but they are hopeful that 'a strong statement that is widely supported can unify and inspire the field.'

9.5.6 In Sum

All of the approaches discussed above contribute to promoting a stronger understanding of research ethics and integrity in students. We do not want to suggest that one of these is better than another. We merely want to argue that they all fulfill their own role in ethics and integrity education, and that universities and educators alike have a never-ending obligation to prepare students the best they can, so that they can prepare the next generation the best *they* can.

9.6 Conclusions

9.6.1 Summary

In this chapter, we examined research ethics in the social sciences from the perspective of a *systems approach*. We situated universities in the dynamic interplay of political and economic forces, and, more specifically, we discussed the influence of new public management politics on university policies. This led us to probe whether there truly is a research misconduct 'crisis' going on in the social sciences.

First, we investigated whether the social sciences suffered from a *replication crisis* – the problem that the few studies that have even been reproduced often fail to be replicated. Along those lines, we also observed that many studies suffer from *weak research rigor* where true findings cannot be distinguished from *false positives*.

Second, we discussed the impact of *perverse incentives* on scientific practices. We found that contrary to what is often suggested, these incentives may not incite fraudulent behavior directly. But there is evidence that they link to several other undesirable trends, including a trend towards *adjunctification of higher education*, or increased prevalence of educators on temporary assignments, which has an indirect impact on research ethics.

Finally, we discussed three different approaches for teaching research ethics and integrity in universities. These approaches can be used to help students come to grips with ethical questions from a *reactive, proactive,* and *reflexive* point of view.

9.6.2   Discussion

This chapter addresses some of the more fundamental problems in the social sciences, specifically those relating to the political and economic forces at play. These forces have resulted in a 'crisis' of sorts, and a drastic restructuring of universities. However, we do not offer political or economic solutions to these issues, which would fall outside the scope of this book.

The second question we did not address is whether teaching ethics and integrity can help solve some of these problems, or could possibly even contribute to them. Some argue that research ethics in the social sciences needs to be modelled after similar, standing practices in the medical sciences, and that governing bodies, such as IRBs, will have to play a more prominent role in upholding professional standards. Others resist this idea based on the objection that a highly professionalized scientific enterprise undermines scientific freedom and creativity (see Resnik 1998, p. 177). There are others still that argue that the formalization of ethical procedures achieves the opposite of what they aim to achieve. Increasingly formalized research ethics structures cause a rupture in the relationship between 'following rules' and 'acting ethically,' and the result of which is called *ethics creep* (Haggerty 2004).

Let us conclude by asking what are your thoughts on this question. Is your institution doing enough, or perhaps even too much, with regard to research ethics? Do you feel prepared to tackle the questions discussed in this book?

Case Study: How to Start a Fire

'Education is not the filling of a pail, but the lighting of a fire'.

In a March 24th, 2017 article in *Psychology Today*, Jesse Marczyk explained how he tried to motivate his students to always identify how they can improve. To achieve this end, he reported to using 'a unique kind of assessment policy,' allowing revisions after grades were received (Marczyk 2017).

From an educational perspective, this seemed 'the reasonable thing to do,' but from a professional perspective, it's plain 'stupid.' When declaring why, Marczyk noted he spent far too much time reading, commenting, and grading papers. For every 100 students, he needed 8 to 16 hours per test, time he argued should be better spent writing grant applications. That is, if he wanted to apply for a tenured track, because hiring committees 'aren't all that concerned with my students' learning outcomes.'

Marczyk's article drives home the point that the commodification and underappreciation of education is directly linked to sciences' replication crisis, and that perverse incentives incite academics to behave in irresponsible ways. When teachers are evaluated on the number of students that pass their tests, they will start 'teaching to the test.' 'Rather than being taught, say, chemistry, *students begin to get taught how to take a chemistry test*, and the two are decidedly not the same thing.'

In this case, we shall look at the consequences of employing a 'technical' perspective on teaching in academic education that emphasizes its instrumental use only (teaching should serve well defined and measurable goals as efficiently as possible). We briefly discuss three authors who criticized, each in their own ways, the assumptions underlying such a perspective.

We begin with Gert Biesta (2007), who offers a series of philosophical arguments against the idea that educational practices should be modeled after the evidence-based approaches found in the field of medicine. Some educationalists, Biesta write, believe that much like the practices in medicine, teachers should rely on evidence from research about what 'works' in education, and establish similar 'evidence-based educationalist practices.' Biesta disagrees with these educationalists.

Underlying their 'evidence-based philosophy' are several untenable assumptions, Biesta argues. First, that there exists 'neutral frameworks' that allow for the assessment of outcomes (namely what is 'effective' in education). In reality, what 'works' in education is not something someone can establish beforehand objectively, it is something that is constructed in the process by teachers and students together.

In a similar vein, James Kennedy (2016), who adheres to Biesta's critique, argues against the idea that effective learning at universities can be both measured and guaranteed. On the contrary, says Kennedy. Learning is not 'a given,' nor is teaching a science – it is an art (2016, p. 35).

But there is room for improvement at universities, Kennedy acknowledges. One suggested area relates to the focus in recent years toward articulating the social impact of research. Universities should strive to better articulate the social impact of teaching and embed it into academic curricula. This would entail providing more room for student engagement, linking their inner concerns and passions to the needs of the world. Kennedy advocates for a new emphasis on student motivation, and a shift towards learner-led programs, that allow students more control over their learning processes.

Finally, we discuss a study by Andrew Boocock (2013), who interviewed administrators, departments heads, and lecturers at a UK college about their experiences with 'performance indicators' (statistics which compare universities and colleges against benchmarks for a number of preestablished, politically motivated outcomes). In this case, the desired outcomes were as follows: (a) better retention rates of students in underrepresented groups, and (b) a higher percentage of students continuing higher education. Certain measures were taken to achieve these goals. How did these indicators impact the motivation of the actors involved?

Administrators often saw great advantages in striving toward performance indicator benchmarks. Student retention rates improved, their achievements went up, guidance and support systems for students were used more efficiently, as were their learning plans.

Teachers perceived a different reality, however. They noticed several side effects of focusing on performance indicators. Student attendance rates and punctuality worsened, which they attributed to the introduction of inertial, bureaucratic disciplinary procedures (Boocock 2013, p. 313). Teachers also experienced a perception of powerlessness and loss of authority, with one teacher describing it as the 'de-professionalized status of lecturers.' Their experience was troubling indeed. 'You're also under pressure in that if you are underperforming they can close your course or put it on special measures internally … We are very much ruled by numbers so that very much changes the tenure of how people behave and how they react' (p. 320).

Teachers furthermore observed perverse incentives, such as to the desire to withdraw 'risky' students and retain 'marginal' students until the end of the academic year (to maximize funding units). All this effected their intrinsic motivation to educate. Even administrators felt the pressure: 'We're now in a position where we're so regulated – we get inspection, external verification, audit, all the bodies coming in and the pressure is felt throughout the organization both on the curriculum and non-curriculum sides' (p. 321).

Assignment

1. Consider your own position in your university. Do you feel that you have enough freedom to develop autonomously as a student? Have you perceived of any 'perverse incentives' in your educational program?
2. Do you agree with Kennedy – that universities should strive to emphasize the social impact of education? What would engage you?
3. Start a fire.

Suggested Reading

For an orientation on the relationship between universities, politics, and economics, we recommend *Through a Darkly Glass: The Social Sciences Look at the Neoliberal University* (Magaret Thornton 2015). *Psychology in Crisis* by psychologist Brian Hughes (2018) grapples with some of the most fundamental problems in his field: theoretical sectarianism, psychology's susceptibility to produce irreproducible results, and convenient sampling, among others. The *Open Science Collaboration* project (2015) is a must read for anyone interested in the replication crisis.

References

Barnard-Brak, L., Saxon, T. F., & Johnson, H. (2011). Publication productivity among doctoral graduates of educational psychology programs at research universities before and after the year 2000. *Educational Psychology Review, 23*(1), 65–73. https://doi.org/10.1007/s10648-010-9146-3.

Bornmann, L., Leydesdorff, L., & Van den Besselaar, P. (2010). A meta-evaluation of scientific research proposals: Different ways of comparing rejected to awarded applications. *Journal of Informetrics, 4*(3), 211–220. https://doi.org/10.1016/j.joi.2009.10.004.

Bouter, L. M. (2015). Commentary: Perverse incentives or rotten apples? *Accountability in Research, 22*(3), 148–161. https://doi.org/10.1080/08989621.2014.950253.

Brall, C., Maeckelberghe, E., Porz, R., Makhoul, J., & Schröder-Bäck, P. (2017). Research ethics 2.0: New perspectives on norms, values, and integrity in genomic research in times of even scarcer resources. *Public Health Genomics, 20*(1), 27–35. https://doi.org/10.1159/000462960.

Canary, H. E. (2007). Teaching ethics in communication courses: An investigation of instructional methods, course foci, and student outcomes. *Communication Education, 56*(2), 193–208. https://doi.org/10.1080/03634520601113660.

Ceci, S. J., Ginther, D. K., & Kahn, S. (2014). Women in academic science: A changing landscape. *Psychological Science in the Public Interest, 15*(3), 75–141. https://doi.org/10.1177/1529100614541236.

Chambers, C. D. (2017). *The seven deadly sins of psychology: A manifesto for reforming the culture of scientific practice*. Princeton: Princeton University Press.

Curnalia, R. M. L., & Mermer, D. (2018). Renewing our commitment to tenure, academic freedom, and shared governance to navigate challenges in higher education. *Review of Communication, 18*(2), 129–139. https://doi.org/10.1080/15358593.2018.1438645.

Dal-Ré, R., Bouter, L. M., Cuijpers, P., Gluud, C., & Holm, S. (2020). Should research misconduct be criminalized? *Research Ethics*, 1–12. https://doi.org/10.1177/1747016119898400.

Dobbie, D., & Robinson, I. (2008). Reorganizing higher education in the United States and Canada: The erosion of tenure and the unionization of contingent faculty. *Labor Studies Journal, 33*(2), 117–140. https://doi.org/10.1177/0160449X07301241.

Esposito, A. (2012). Research ethics in emerging forms of online learning: Issues arising from a hypothetical study of MOOCS. *The Electrotonic Journal of e-Learning, 10*(3), 315–325. http://academic-conferences.org/ejournals.htm.

Fanelli, D. (2010a). "Positive" results increase down the hierarchy of the sciences. *PLoS One, 5*(4), e10068. https://doi.org/10.1371/journal.pone.0010068.

Fanelli, D. (2010b). Do pressures to publish increase scientists' bias? An empirical support from US states data. *PLoS One, 5*(4), e10271. https://doi.org/10.1371/journal.pone.0010271.

Fanelli, D., & Larivière, V. (2016). Researchers' individual publication rate has not increased in a century. *PLoS One, 11*(3), e0149504. https://doi.org/10.1371/journal.pone.0149504.

Finkelstein, M. (2014). The balance between teaching and research in the work life of American academics. In J. Shin, A. Arimoto, W. Cummings, & U. Teichler (Eds.), *Teaching and research in contemporary higher education. The changing academy – The changing academic profession in international comparative perspective, vol 9* (pp. 229–318). Dordrecht: Springer. https://doi.org/10.1007/978-94-007-6830-7_16.

Garvin, D. (1980). *The economics of university behavior*. New York: Academic.

Gelstein, S., Yeshurun, Y., Rozenkrantz, L., Shushan, S., Frumin, I., Roth, Y., & Sobel, N. (2011). Human tears contain a chemosignal. *Science, 331*(6014), 226–230. https://doi.org/10.1126/science.1198331.

Ginther, D. K., & Kahn, S. (2009). Does science promote women? Evidence from academia 1973-2001. In *Science and engineering. Careers in the United States: An analysis of markets and employment* (pp. 163–194). Chicago: The University of Chicago Press.

Gračanin, A., Bylsma, L. M., & Vingerhoets, A. J. J. (2014). Is crying a self-soothing behavior? *Frontiers in Psychology, 5*, 502. https://doi.org/10.3389/fpsyg.2014.00502.

Gračanin, A., van Assen, M. A. L. M., Omrčen, V., Koraj, I., & Vingerhoets, A. J. M. (2017a). Chemosignalling effects of human tears revisited: Does exposure to female tears decrease males' perception of female sexual attractiveness? *Cognition and Emotion, 31*(1), 139–150. https://doi.org/10.1080/02699931.2016.1151402.

Gračanin, A., Vingerhoets, A., & Assen, M. A. (2017b). Response to comment on "Chemosignalling effects of human tears revisited: Does exposure to female tears decrease males' perception of female sexual attractiveness?". *Cognition & Emotion, 31*(1), 158–159. https://doi.org/10.1098/rstb.2019-0262.

Haggerty, K. D. (2004). Ethics creep: Governing social science research in the name of ethics. *Qualitative Sociology, 27*(4), 391–414. https://doi.org/10.1023/B:QUAS.0000049239.15922.a3.

Hangel, N., & Schmidt-Pfister, D. (2017). Why do you publish? On the tensions between generating scientific knowledge and publication pressure. *Aslib Journal of Information Management, 69*(5), 529–544. https://doi.org/10.1108/AJIM-01-2017-0019.

Herbst, M. (2007). Alternative governance and management modes. In *Financing public universities. Higher Education Dynamics* (Vol. 18). Dordrecht: Springer. https://doi.org/10.1007/978-1-4020-5560-7_5.

Hicks, D. (2012). Performance-based university research funding systems. *Research Policy, 42*(2), 251–261. https://doi.org/10.1016/j.respol.2011.09.007.

Hood, C. (1991). A public management for all seasons? *Public Administration, 69*(1), 3–19. https://doi.org/10.1111/j.1467-9299.1991.tb00779.x.

Hughes, B. M. (2018). *Psychology in crisis*. London: Macmillan.

Huyer, S. (2015). Is the gender gap narrowing in science and engineering? In *UNESCO science report: Towards 2030* (pp. 85–104). Paris: UNESCO publishing.

Karkowski, A. M. (2010). Activities to help students appreciate committees that protect research participants. *Council on Undergraduate Research Quarterly, 30*(3), 11–15.

Kezar, A., Maxey, D., & Holcombe, E. (2015). The professoriate reconsidered a study of new faculty model. The Delphi project on the changing faculty and student success, University of Southern California.

Kuhn, T. S. (1970). *The structure of scientific revolutions*. Chicago: Chicago University Press.

Löfström, E. (2016). Role-playing institutional academic integrity policy-making: Using researched perspectives to develop pedagogy. *International Journal for Educational Integrity, 12*(1), 5. https://doi.org/10.1007/s40979-016-0011-0.

Makel, M. C. (2017). The empirical march: Making science better at self-correction. *Psychology of Aesthetics Creativity and the Arts, 8*(1), 2–7. https://psycnet.apa.org/doi/10.1037/14805-037.

Makel, M. C., Plucker, J. A., & Hegarty, B. (2012). Replications in psychology research. *Perspectives on Psychological Science, 7*(6), 537–542.

Maturena, H., & Varela, F. (1980). *Autopoiesis and cognition: The realization of the living (Vol. 42)*. Springer Science & Business Media.

Maxwell, S. E., Lau, M. Y., & Howard, G. S. (2015). Is psychology suffering from a replication crisis? What does "failure to replicate" really mean? *American Psychologist, 70*(6), 487–498. https://doi.org/10.1037/a0039400.

Merton, R. K. (1968). The Matthew effect in science: The reward and communication systems of science are considered. *Science, 159*(3810), 56–63. https://doi.org/10.1126/science.159.3810.56.

Merton, R. K. (1988). The Matthew effect in science, II: Cumulative advantage and the symbolism of intellectual property. *Isis, 79*(4), 606–623. https://doi.org/10.1086/354848.

Moustafa, K. (2014). The disaster of the impact factor. *Science and Engineering Ethics, 21*(1), 139–142. https://doi.org/10.1007/s11948-014-9517-0.

Naimi, L. (2007). Strategies for teaching research ethics in business, management and organisational studies. *The Electronic Journal of Business Research Methods, 5*(1), 29–36.

Ogbogu, C. O. (2011). Gender inequality in academia: Evidences from Nigeria. *Contemporary Issues in Education Research, 4*(9), 1–8. Retrieved from: https://files.eric.ed.gov/fulltext/EJ1072918.pdf.

Open Science Collaboration. (2015). *Science*, https://doi.org/10.1126/science.aac4716

Parson, T. (1951). *The social system*. Glencoe: The Free Press.

Resnik, D. B. (1998). *The ethics of science: An introduction*. London/New York: Routledge.

Schwartz, J. M. (2014). Resisting the exploitation of contingent faculty labor in the neoliberal university: The challenge of building solidarity between tenured and non-tenured faculty. *New Political Science, 36*(4), 504–522. https://doi.org/10.1080/07393148.2014.954803.

Siegel, D., & Baveye, P. (2010). Battling the paper glut. *Science, 329*(5998), 1466. https://doi.org/10.1126/science.329.5998.1466-a.

Simmons, J. P., Nelson, L. D., & Simonsohn, U. (2011). False-positive psychology: Undisclosed flexibility in data collection and analysis allows presenting anything as significant. *Psychological Science, 22*(11), 1359–1366. https://doi.org/10.1177/956797611417632.

Smith, N. C. (1970). Replication studies: A neglected aspect of psychological research. *American Psychologist, 25*(10), 970–975. https://psycnet.apa.org/doi/10.1037/h0029774.

Sobel, N. (2017). Revisiting the revisit: Added evidence for a social chemosignal in human emotional tears. *Cognition & Emotion, 31*(1), 151–157. https://doi.org/10.1080/02699931.2016.1177488.

Sponholz, G. (2000). Teaching scientific integrity and research ethics. *Forensic Science International, 113*(1–3), 511–514.

Stern, J. E., & Elliott, D. (1997). *The ethics of scientific research: A guidebook for course development*. Hanover/London: University Press of New England.

Stone, D. (2002). *Policy paradox: The art of political decision making*. New York: W.W. Norton & Company.

Sweet, S. (1999). Using a mock institutional review board to teach ethics in sociological research. *Teaching Sociology, 27*(1), 55–59. https://www.jstor.org/stable/1319246.

Thornton, M. (2015). *Through a darkly glass: The social sciences look at the neoliberal university*. Acton: ANU Press.

Van Bavel, J. J., Mende-Siedlecki, P., Brady, W. J., & Reinero, D. A. (2016). Context sensitivity in scientific reproducibility. *Proceedings of the National Academy of Sciences, 113*(23), 6454–6459. https://doi.org/10.1073/pnas.1521897113.

van den Besselaar, P., & Leydesdorff, L. (2009). Past performance, peer review and project selection: A case study in the social and behavioral sciences. *Research Evaluation, 18*(4), 273–288. https://doi.org/10.3152/095820209X475360.

Van Wesel, M. (2016). Evaluation by citation: Trends in publication behavior, evaluation criteria, and the strive for high impact publications. *Science and Engineering Ethics, 22*(1), 199–225. https://doi.org/10.1007/s11948-015-9638-0.

von Unger, H. (2016). Reflexivity beyond regulations. Teaching research ethics and qualitative methods in Germany. *Qualitative Inquiry, 22*(2), 87–98.

Wiener, H. (1965). *Cybernetics, second edition: Or the control and communication in the animal and the machine*. Cambridge: The MIT Press.

References for Case Study: How to Start a Fire

Biesta, G. (2007). Why "what works" won't work: evidence-based practice and the democratic deficit in educational research. *Educational Theory, 57*(1), 1–22. https://doi.org/10.1111/j.1741-5446.2006.00241.x.

Boocock, A. (2013). Further education performance indicators: a motivational or a performative tool? *Research in Post-Compulsory Education, 18*(3), 309–325. (2013. https://doi.org/10.1080/13596748.2013.819272.

Kennedy, J. (2016). An educated guess. In M. Flikkema (Ed.), *Sense of serving. Reconsidering the role of universities today* (pp. 32–41). Amsterdam: University Press.

Marczyk, J. (2017, March). Academic perversion. *Psychology Today*. Retrieved from https://www.psychologytoday.com/us/blog/pop-psych/201703/academic-perversion

Part IV
Forms, Codes, and Types
of Regulations

Chapter 10
Research Ethics Step by Step

Contents

This chapter has been co-authored by Dorota Lepianka.

Electronic Supplementary Material: The online version of this chapter (https://doi.org/10.1007/978-3-030-48415-6_10) contains supplementary material, which is available to authorized users.

227

J. Bos, *Research Ethics for Students in the Social Sciences*,
https://doi.org/10.1007/978-3-030-48415-6_10

After Reading This Chapter, You Will:

- Have a general knowledge of Institutional Review Board (IRB) procedures
- Have the capacity to anticipate the basic ethical pitfalls in research designs
- Know how to counter common ethical objections
- Be able to design an informed consent form

Keywords Added value · Anonymization · Avoiding harm · Benefits · Conflicting loyalties · Cost-benefit analysis · Data management · Data storage · Deception · Doing good · Equitability · Ethics creep · Expenditures · False negatives · False positives · Gatekeepers · Informed consent · Intrusive questioning · Invasion of integrity · IRB · Loyalty · Privacy · Protocols · Pseudonymization · Reciprocity · Relevancy · Responsibility · Risk assessment · Seeking justice · Trust · Vulnerability

10.1   Introduction

10.1.1   Research Design and Ethical Approval

In this chapter, we aim to guide you through some of the most important ethical issues you may encounter throughout the process of finalizing your research design and preparing it for the process of ethical approval. The issues discussed here range from broad topics about the relevancy of the research itself, to detailed questions regarding confidentiality, establishing informed consent, briefing and debriefing research participants, dealing with invasive techniques, deception, and safe storage of your data.

The majority of these ethical dilemmas coincide largely with the concerns voiced by independent *Institutional Review Boards* (IRBs, also referred to as *Research Ethics Committees*, or RECs). IRBs register, review, and oversee local research applications that involve human participants. They are established to protect the rights of research participants and to foster a sustainable research environment. The task of such boards is to evaluate whether or not a research design meets the institutional ethics standards and facilitates a necessary *risk assessment*.

The necessity of ethical reviewing is reflected in national laws as well as international declarations and has become a mandatory procedure in universities and research institutes worldwide (see Israel 2015, for an overview of ethical reviewing practices). Failing to seek the approval of an IRB can have serious consequence for the researchers involved. For example, the retraction note attached to an article on bullying, published in 2017 in the *International Journal of Pediatrics* revealed 'The study was conducted in agreement with the school principal and the authors received verbal approval, but they did not receive formal ethical approval from the designated committee of the Ministry of Education' (entry at 'Retraction Watch', March 13th 2019).

A number of scholars focusing on ethical review processes have critiqued the institutionalization of ethical reviewing, because, as one author observed, it seems to assume that unscrupulous researchers are restrained only by the leash and muzzle of the IRB system (Schneider 2015, p. 6).

Indeed, by setting aside ethics as a separate issue and submitting it to an 'administrative logic' (procedural, formalistic approach), scholarly research has fallen prey to a form of *ethics creep,* a process whereby the regulatory system expands and intensifies at the expense of genuine ethical reflection (Haggerty 2004). Scott (2017) remembers how a simple study once was killed by such formalistic procedures. Understandably, researchers sometimes see the completion of an IRB application form to be a mere 'formality, a hurdle to surmount to get on and do the research' (Guillemin and Gillam 2004, p. 263).

We agree that ethical considerations should inform our discussions about research, and that these discussions should not be obstructed by regulatory procedures. The aim of this chapter is therefore to assist you in your ethical deliberations. This chapter seeks to guide you through the process of making important ethical decisions at all stages of formulating a research design, and to help you identify the common pitfalls, objections, and critiques. To facilitate this process, we have designed a series of queries at the end of each paragraph, that could be taken into consideration whenever you plan to carry out a research project. Not all questions may be relevant to all research projects, but as a whole, they should facilitate a fairly thorough preparation.

In the sections to follow, we map out the various ethical dimensions of designing a research project step by step: addressing the fundamental question of why and for whom we do research (Sect. 10.2); an exploration of the ethical considerations of the research design itself, including the recruitment of study participants (Sects. 10.3 and 10.4); violation of integrity (Sect. 10.5); avoiding deception (Sect. 10.6); informed consent (Sect. 10.7); collecting data during field work (Sect. 10.8); what to do with incidental findings (Sect. 10.9); analyzing collected data (Sect. 10.10); reporting and dissemination of research findings (Sect. 10.11); and finally data management and storage (Sect. 10.12). This chapter closes with a summary (Sect. 10.13) and we include a brief ethics checklist and offer a model informed consent form that can be used in the future to help you cross all your 't's and dot all your 'i's (Box 10.1).

We highlight our discussions with multiple case studies selected from a wide range of disciplines within the social sciences, including specializations within psychology, anthropology, educational sciences, interdisciplinary studies, and others. For the sake of brevity, we refrain from seeking examples from all disciplines for each individual dilemma, but instead focus on those that seem most poignant. We hope this overview will prepare you to face the rigors of research with confidence.

10.2   Relevancy: Choice of Research Area

10.2.1   What for?

There are few subjects or questions that researchers cannot study, but are they all worth researching? That is a different question. Contrary to what you may think, completely new research questions do not exist. Research builds upon the pre-existing research lexicon. In fact, researchers have an obligation to enhance or critique theories, improve established bodies of knowledge, and adapt or alter relevant methodologies.

Failing to acknowledge research traditions may come with the risk of wasting valuable resources, but also of self-disqualification. The relevancy of a research project is thus not so much measured in terms of how *much* knowledge it generates, but rather in how much knowledge it generates *in relation to what is already known* (see the imperative of originality, discussed in Chap. 2).

Box 10.1: Rules of Thumb for Ethical Assessment of Research Designs

1. **Avoiding Harm** Researchers have a responsibility to ensure that their study does no harm to any participants or communities involved. They also need to assess the risks that participants (and communities) may face.

    How likely is your research project to cause harm to the individuals or communities you choose to research? How serious is the possible harm? What measures need be taken to offset the risks? Is there any way in which harm could be justified or excused? How do you ensure that your study does not endanger the values, cultural traditions, and practices of the community you study?

2. **Doing Good** Researchers have the complementary obligation to do research that contributes to the furthering of others' well-being.

    Who are the beneficiaries of your study? What specific benevolence might flow from it and for whom? What can participants reasonably expect in return from you and what should you offer them, if anything? What does your study offer to promote the well-being of others? How does the community or society at large benefit?

3. **Seeking Justice** Finally, researchers should ensure that participants are treated justly and that no one has been favored or discriminated against.

    Do you treat your participants fairly and have you taken their needs into consideration? How do you ensure a fair distribution of the burdens and benefits in both the participant's experience and research outcomes? How are the (perhaps contradictory) needs of the communities taken into account?

    Whereas all three criteria seem 'self-evident' if not trivial, there remains the critical and difficult question of *how* to interpret them, and whether they apply in any given case (i.e. everybody will agree that one should not harm people and do good or seek justice but what does this mean in practice?). For further discussion, see Beauchamp and Childress (2001), *Principles of Biomedical Ethics*.

10.2.2 For Whom?

Some research is fundamental – for the sake of knowledge – but most is not. Often, results have certain practical uses for other parties, sciences' *stakeholders*. They can be *commissionaires* who act as patrons of research projects, *professionals* working in a 'field of practice' who make use of scientific knowledge, or their *clients*. Research can have implications for policy makers, teachers, therapists, professionals working with minority groups, or indeed, minority groups themselves, to name but a few.

The question how research projects impact various (potential) stakeholders is not always explicitly addressed, but we feel that this is something that deserves careful attention. Who is addressed, who will be influenced, and who can make use of research in which ways? Consider the following two examples where the stakeholders are specifically targeted and even addressed.

1. Ran et al. (2003) describe a comparative research study into the effectiveness of psychoeducational intervention programs in the treating of schizophrenics in rural China. The program specifically targets patients' relatives, who, the researchers conclude, need to improve their knowledge of the illness and change their attitude towards the patient.
2. A qualitative study on experiences with prejudice and discrimination among Afghan and Iranian immigrant youth in Canada singles out the media as a 'major contributor to shaping prejudicial attitudes and behaviors,' and schools as one of the first places youth may encounter discrimination (Khanlou et al. 2008)

10.2.3 At What Cost?

Thirdly, there is the question of balancing costs and benefits of research. *Costs* comprise of salaries, investments, use of equipment, but also of sacrifices or (health) risks run by all those involved. *Benefits* can be expected revenue and earnings, but also gained knowledge and expertise, certain privileges allotted to participants, or even access to particular facilities.

The fact that the costs and benefits can be of a material and immaterial nature makes them both difficult to measure and predict (see Diener and Crandall 1978). How do you value and weigh costs and benefits? Who should profit and who should run which risks?

While there is no way to answer these questions in general, there are different models that you can use to assess risks and benefits, based on what you think counts as important.

In the first model, science is committed to the *principle of impartiality*. Researchers and research participants partake in research primarily because they value science, want to promote its cause, and feel that their contribution helps further scientific knowledge. In this model, costs consist just of the salaries of the researchers and the marginal compensation of the participants for their time. Knowledge acquisition is the most important gain, and risks are understood in the immediate context of research (health hazards).

In the second model, knowledge production is regarded as a commercial activity. Universities and their researchers are seen as entrepreneurs who collaborate with other parties (mainly industry and government) and are committed to the *principle of profit*. In this model, costs are seen as investments, gains as (potential) revenues. Compensation of participants is an expense item and any risks they run can be 'bought off.'

The third model proposes knowledge production from the *principle of equitability* (fairness for all). It accepts that knowledge may be profitable, but rejects a one-sided distribution of gains, where all the profits (patents, publication, prestige, grants) go to the researchers only, and none to the participants. Participants should not merely be monetarily compensated, but profit in a much more direct way, for example by giving them access to health facilities, providing better knowledge of the topic in question (Anderson 2019) or even empowering whole communities (Benatar 2002).

These different models not only perceive parties or stakeholders differently, they also perceive of risks, costs and benefits differently. Consequently, researchers may come to weigh the costs, benefits, and risks differently depending on what they value most (Box 10.2).

10.2.4   Trauma Research: A Case in Point

Consider the question of whether research on traumatic experiences itself should be regarded as harmful. It is argued, on the one hand, that asking about traumatic experience is risky, as survivors may be more vulnerable and easier to stigmatize. On the other hand, there is also evidence that suggests that talking or writing about traumatic experiences can in fact be beneficial, psychologically as well as physiologically (Marshall et al. 2001). How does one weigh the (potential) risks against the (possible) benefits (DePrince and Freyd 2006)?

Box 10.2: Fair Compensation?
In a research application for a study on coping with undesirable social behavior at the workplace in China, the researchers planned to ask participants to complete a questionnaire which was estimated to takes up to 15 min. Participants would receive ¥8 (roughly 1 Euro) in compensation for their effort, but only once they completed the questionnaire. When queried by their local IRB why every participant wouldn't be compensated regardless, rather than only those who complete the questionnaire, the researchers presented four arguments:

1. Rewarding participation before finishing the research leads to high drop-out rates
2. It is difficult to organize payment with non-completers
3. The questions are non-invasive
4. In comparable cases, applications are always approved by IRBs

Evaluate these responses by ranking the arguments. Which argument do you find most and which least convincing, and why?

(Case communicated to one of the authors of this chapter)

In a study among 517 undergraduate students, Marno Cromer et al. (2006) asked subjects to rate how distressing it was for them to discuss a range of traumatic experiences and found that a vast majority did not find it difficult at all. However, argued the authors sensibly, it's not the *average* that counts here, but the *exception*. And indeed, 24 participants reported the trauma research to be 'much more distressing' than everyday life. Of these 24, all but one still believed the research to be important enough to be carried out. The one exception reported that the research seemed 'a somewhat bad idea.'

These findings concur with Newman et al. (1999), who did research on childhood abuse and found that a minority of the respondents reported feeling upset after the research. Of these, a few indicated that they would have preferred to have not participated had they known what the experience would be like.

In weighing the (immaterial) benefits against the costs of talking about traumatic experiences (distress), the former were deemed to outweigh the latter, provided that interviewers are carefully selected and trained.

Of note, however, one consideration is left out of this comparison, namely the question of whether *not* doing the research should be considered a risk (Box 10.3). Indeed, Becker-Blease and Freyd (2006, p. 225) reason that 'silence is part of the problem', and there is a real 'possibility that the social forces that keep so many people silent about abuse play out in the institution, research labs, and IRBs.'

Will the cost-benefit balance shift if the risk of *not* doing research be taken into consideration?

Box 10.3: Risks and Benefits

Risk: The probability of harm (physical, psychological, social, legal, or economic) occurring as a result of participation in a research study. The probability and the magnitude of possible harm may vary from minimal (or none) to significant.

Minimal Risk: A risk is considered to be minimal when the probability and magnitude of harm or discomfort anticipated in the proposed research are not greater, in and of themselves, than those ordinarily encountered in daily life or during the performance of routine physical or psychological examination or tests.

Benefit: A valued or desired outcome, of material or physical nature (i.e. money, goods), or immaterial nature (i.e. knowledge, skills, privileges). Individuals may not only benefit from the research, but also communities as a whole.

(Adapted from the Policy Manual of the University of Louisville.)

Q1: What is the added value of my research project and for whom does it benefit?

• Which research traditions and methodologies do I relate to and why?
• Who is addressed by my research project (who are my possible stakeholders)?
• Which costs and benefits can be expected, for whom, and how do I balance them?

10.3 Choice of Participants

10.3.1 Ethical Limitations in Choice of Participants

Researchers must make many decisions regarding the choice of participants. Is the sample randomly selected and does it give a fair representation of the population? Will the N be large enough to test my hypothesis? Has non-response been taken into account? Et cetera. Some of these methodological questions have ethical consequences, as we will explore below.

10.3.2 Number of Participants

This is of ethical concern because research is considered (at least to a degree) a burden on participants and often times on society at large as well. The number of participants should therefore be no more than absolutely necessary.

In quantitative studies, a reasonable estimate can be given with a *power analysis*. 'Statistical power' in hypothesis testing signifies the probability that the test will detect an effect that actually exists. By calculating the power of a study, it becomes possible to determine the required sample size, given a particular statistical method, and a predetermined degree of confidence. For example, to detect a small interaction effect between two variables, using a linear mixed-effect method, a sample of $N = 120$ would suffice at a default alpha of .05. Remaining space in this book does not permit a detailed discussion of how to calculate the power of a study, but see Cohen 1988, for an explanation of power in the behavioral sciences in general.

In qualitative studies, no such power analysis would be suitable. Instead, the principle of *saturation* is often used. Saturation implies approaching new informants until enough knowledge is gained to answer the research question, or until the categories used are fully accounted for. What exactly constitutes 'saturation' may differ from one field of expertise to the next and may need further problematization moving forward (see O'Reilly and Parker 2012).

10.3.3   Selection of Participants

Laboratory studies often use undergraduate students as research subjects (usually in exchange for 'credits'). These are called *subject pools*. In some fields of research in the social sciences, subject pools make up the majority of research participants, as Diener and Crandall (1978) pointed out long ago.

Convenience sampling (using groups of people who are easy to contact or to reach) not only has methodological drawbacks, but also ethical implications. Heinrich, Heine, and Norenzayan (2010) called attention to the social science's 'usual subjects' and named them WEIRDOs, an acronym for Western, Educated, Industrialized, Rich and coming from Democratic cultures.

They maintain that WEIRDOs aren't representative of humans as a whole, and that psychologists shouldn't routinely use them to make broad claims about the drivers of human behavior because WEIRDOs differ in fundamental aspects with non-WEIRDOs. Different cultural experiences result in differing styles of reasoning, conceptions of the self, notions of fairness, and even visual perception.

10.3.4   Online Communities

As a specific target group for research, online communities pose their own dilemmas. Legally, researchers must be aware that they may be bound by the 'general terms and conditions' of these online platforms, which can restrict the use of their data for research purposes. Morally, it is important to ask whether it is right to record the activities of an online public place without the participant's consent, regardless of whether it is allowed (see Chap. 7 for a discussion of this question). There are two viewpoints we will explore on this matter.

Oliver (2010, p. 133) argues that although communication in an online environment may be mediated in different ways, it is still communication between people. In essence, the same ethical principles should apply, including the receipt of active consent.

Burbules (2009, p. 538) on the other hand, argues that in online or web-based research, notions regarding privacy, anonymity, and the right to 'own' information needs to be radically reconsidered.

What matters online, Burbules argues, is not so much anonymity, but rather access. In the digital universe, people *want* to share information. But they also want to *control* who can make use of it. A challenging dichotomy to navigate indeed.

This problem (the question of who can access which data) has become even more urgent today. This urgency can be traced to new information and communication technologies that enable researchers to build extremely complex models based on massive and diverse databases, allowing increasingly accurate predictions about an individual's actions and choices.

10.3.5 Control Groups

Research on the effects of certain interventions that involve *control groups* (participants who receive either less effective or no treatment) leads to the question of whether it is fair for a participant to be placed in a disadvantageous position.

This is referred to as *asymmetrical treatment*. The question is grounded in considerations of *egalitarian justice*, which is in other words, the idea that individuals should have an equal share of the benefits, rather than just the baseline avoidance of harm.

It is suggested that participants in the control group be offered the more effective treatment once the study is completed (Mark and Lenz-Watson 2011). A problem with this being that it applies to certain research designs only (typically RCT, or 'Randomized Controlled Trial') and not to others (policy interventions, for example, or education; for further discussion, see Diener and Crandall 1978). With research on policy interventions, (as opposed to treatment research), the question is whether or not it is fair to offer certain policies to certain groups and not to others.

Q2: Who are the participants in my research project?

- Which ethical consequences may be involved in selecting participants for my research project?
- How do I ensure that my selection of participants does not result in unfair treatment?

10.4 Vulnerable Participants

10.4.1 Vulnerable Participants

Vulnerable participants are properly conceived of as those who have 'an identifiably increased likelihood of incurring additional or greater wrong' (Hurst quoted in Bracken-Roche et al. 2017). Seeking the cooperation of vulnerable people may be problematic for various reasons, but that does not imply that they cannot or should not be involved in research. It does mean that these groups need special attention, however.

10.4.2 Minors and Children

Working with minors and children requires consideration from both a moral and a legal perspective. Often in place are somewhat arbitrary age limits that will differ from country to country, which require that researchers seek active consent from the

238 10 Research Ethics Step by Step

parents or legal representatives of a child. This says little, however, about the minor's moral capacity to participate in research.

IRBs generally acknowledge that children can be involved, but that different age-groups should be treated on par with their stages of psychological development, that will inform what a six-year-old or a twelve-year-old child can or cannot do, or what an eight-year-old is capable of comparatively. In general, the younger the child, the shorter and less intense the inquiry should be.

We concur with Schenk and Rama Rao (2016, p. 451) who argue that young children should be excluded from providing detailed information on potentially traumatic topics that may cause strong emotional distress. As is usually the case, exceptions can be made under particular circumstances, but they remain outliers. We also agree with Vargas and Montaya (2009) that it is sensible to consider any contextual and cultural factors, as this may make a difference in a child's understanding of the research environment.

Finally, we emphasize that researchers who work with minors (children) should have special training on how to interview or collect data from them.

10.4.3 Disadvantaged Participants

When cognitively impaired individuals are included in a research design, special attention must be paid to the potential level of invasiveness, the degree of risk, the potential for benefit, and the participant's severity of cognitive impairment (Szala-Meneok 2009). Likewise, people who are in *dependent circumstances* (such as detainees, elderly people in nursing homes, or the unemployed), may not always have the capacity to refuse consent, or may fail to understand that they have the power to refuse cooperation. A reasonable assessment regarding the perceived ability to participate and to refuse participation must thus be made for every case in which these populations are involved (see Box 10.4 for an overview of vulnerable participants).

10.4.4 Mixed Vulnerability

At times, several forms of vulnerability coincide within one research proposal. Consider as a case in point a proposed study into health problems (suicidal ideation), sexual risk-taking behavior, and substance use of LGBT adolescents of between 16–17 years old, as reported by Brian Mustanski (2011). An Institutional Review Board (IRB) was hesitant to approve Mustanski's application for a number of reasons. We will discuss those reasons below, together with Mustanski's responses.

The first problem the IRB encountered was that the researcher was seeking a waiver for parental permission. Adolescents at this age are legally minors and any

Box 10.4: Vulnerable Populations by Category

Category	Examples (not exhaustive)
Cognitive or communicative Vulnerability participants who are unable to process, understand, or appreciate consent either by mental or language limitation. Researchers targeting a population where this is likely to be present must provide a consent procedure that will accommodate the needs of participants, either by translating documents, writing it in basic language, or discussing the consent.	People with little or no literacy skills People with cognitive impairment
Institutional Vulnerability this includes individuals who are subject to a formal authority and whose consent may be coerced, either directly or indirectly. A solution to this issue can be using a third party to advise on the matter, and possibly eliminate any conflicts of interest.	Prisoners Student/professor relationships Employee/employer relationships
Deferential Vulnerability individuals who informally act as subordinates to an authority figure, where one party may feel obligated to follow the advice of another. These situations require a sensitive recruitment and consent plan where participants have the opportunity to consent voluntarily.	Abuse victims Doctor/patient relationships Husband/wife relationships
Medical Vulnerability this includes individuals with a medical condition that may cloud their ability to make decisions regarding their participation. The patient may see the research study as a miracle cure to their disease instead of a procedure that has no guarantee for results. Researchers should ensure that participants are able to understand the full meaning of the study to alleviate any misunderstanding.	Patients People with incurable diseases Very sick people
Economic Vulnerability this includes individuals whose economic situation may make them vulnerable to the prospect of free care and/or the payments issued for participating in the study. It is important that the payment offered will not encourage an individual to put themselves at a greater risk than they would otherwise.	Homeless people Unemployed people People on welfare benefits or social assistance
Social Vulnerability participants who are at risk for discrimination based on race, gender, ethnicity, or age fit into this category. The participant may also be prone to feel discriminated against and may not participate as a result of this predisposition.	Ethnic minorities LGBTQ Elderly
Legal Vulnerability this includes participants who do not have the legal right to consent or who may be concerned that their consent could put them at risk for legal repercussions. For those who are unable to legally consent, it is important that you obtain consent from a legal representative and in most cases also obtain consent from the individual.	Minors People under legal guardianship
Study Vulnerability participants who are made vulnerable by the study's design, specifically through deception. This can be alleviated by ensuring full consent and disclosure after the study is completed (debriefing) or whenever a participant withdraws from the study.	Any participant who is subjected to deception

Adapted with permission from the guidelines of the Institutional Review Board for Social and Behavioral Sciences at the University of Virginia.

waiver requires the provision of an appropriate mechanism for protecting the minor. Mustanski argued, however, that the goal of waiving parental permission was not to circumvent the authority of parents. 'Instead, it is to allow for scientists to conduct research that could improve the health of adolescents in cases where parental permission is not a reasonable requirement to protect the participating youth' (2011, p. 677).

The second concern of the IRB was the vulnerability of the LGBT community collectively, who have historically been more prone to stigmatization and discrimination. Mustanski replied that he knew of no evidence that demonstrated any decision-making impairment of members of the LGBT community, and that he believed many of them would be insulted to have it implied otherwise.

Finally, and perhaps most importantly, the IRB was worried that participants in this research might be exposed to sensitive information that could lead to psychological harm. Mustanski agreed that IRBs should have a role in protecting participants' interests, but argued that IRBs tended to overestimate risks. This can lead to time-consuming procedures and the implementation of supposed protections that may mitigate the scientific validity of the research, or discourage future behavioral research involving certain populations. After a number of required adjustments (such as a more detailed risk assessment), the proposal was ultimately accepted.

Q3: How do you ensure appropriate and equitable selection of participants?

- Who are your research subjects?
- Are your research subjects part of a vulnerable population, and if so, what risks do you anticipate?
- Where do you expect to find them and how do you intend to recruit them?

10.5   Use of Invasive Techniques

10.5.1   Invasive Techniques

By *invasive techniques*, we mean any procedure or intervention that affects the body or mind of a research participant such that it results in psychological or physical harm. Some argue that our definition of invasiveness should not be limited to individual participants but should include entire communities as well (Box 10.5).

Invasive techniques, by definition, violate the *principle of nonmaleficence* ('do no harm'), and are among the most urgent concerns of IRBs the world over. However, harm is broadly (and vaguely) defined, ranging from trauma to strong disagreeable feelings, and from short-term to long-lasting. *The European Textbook on Ethics* (2010, p. 200) defines harm as such: 'To be harmed is to have one's interests set back or to be made worse off than one would otherwise have been. Harms can relate to any aspect of an individual's welfare, for example physical or social.

Box 10.5: Invasive or Intrusive?

The term *invasive* originates from the medical sciences, where it means: *entering the body, by cutting or inserting instruments*. In the social sciences, it describes techniques that enter one's privacy. Questions about one's sexual orientation, political preferences, and other privately sensitive subjects are considered 'invasive', as is exposure to strong aversive stimuli or traumatizing experiences.

Intrusive was originally a legal term, described as *entering without invitation or welcome*. In the social sciences, it describes techniques that invoke 'unwelcome feelings.' Research may be regarded as 'intrusive' when it concerns topics that respondents dislike talking about or find difficult to discuss (Elam and Fenton 2003, p. 16). Intrusive techniques can also involve prolonged procedures and processes that involve substantial physical contact. Intrusive questions can make a participant feel uneasy, uncomfortable, even shameful: 'Are you anorexic?' 'Do you masturbate?'

Some examples of invasive and intrusive technique include:

- EEG, PAT scans, CAT scans, (f)MRI, or measuring heart rate, are all non-invasive in the medical and psychological sense, but can be intrusive.
- Questions about race, ethnicity, and sexual health can be both invasive and intrusive.
- Queries about personal information, including name, date and place of birth, biometric records, education, financial, and employment history, are often thought to be neither invasive nor intrusive. However, to some people some of these questions *can* be intrusive. Regardless, use of this information is strictly limited under data protection regulation in most countries.

Institutions can also be harmed insofar as they can be thought of as having interests distinct from those of their members.'

Invasive techniques may include exposure to insensitive stimuli, intrusive interrogation, excessive measurements, or any procedures that can cause damage. We exclude from our discussion any medical practices or intervention, such as administering drugs or the use of clinical health trials and refer anyone who intends to use these techniques to specialized Medical Research Ethics Committees (MRECs).

10.5.2   Examples of Invasive Research

Some of psychology's most famous experiments were hampered by the ethical quandaries of invasive research. For example, John Watson's 1919 behavioral experiments with 'Little Albert,' an eleven-month-old child, who was exposed to

Fig. 10.1 *Little Albert.*
Still from the film made by
Watson. (Source:
Wikipedia)

Now he fears even Santa Claus

loud, frightening sounds when presented with specific fearsome images. Although
it is unclear what the net effects were on the child, by today's standards, the design
would be considered unethical for its gross lack on concern for the wellbeing of the
child (see Harris 1979; Beck et al. 2009) (Fig. 10.1).

Harry Harlow's 'Pit of Despair Studies' from the 1950s involved infant primates
who were raised in social isolation, without their protective mothers or with surro-
gate mothers (dolls). They consequently developed signs of what humans call 'panic
disorder.' This complete lack of concern for animal welfare would certainly be considered unethical by today's standards.

Psychologist Stanley Milgram's well-known 1961 experiments, that involved
participants who were led to believe that they were administering electric shocks to
fellow participants are deemed invasive, despite the researcher attempting to minimize harm by debriefing his participants (see Tolich 2014).

Diana Baumrind (1964) was quick to recognize the ethical perils of the Milgram
studies: 'From the subject's point of view procedures which involve loss of dignity,
self-esteem, and trust in rational authority are probably most harmful in the long run
and require the most thoughtfully planned reparations, if engaged in at all' (p. 423).

10.5.3 Avoiding Invasive Routines?

Can (or should) invasiveness be avoided at all times? The answer seems obvious: no
techniques that cause harm should be put to use. In practice, however, the answer is
more ambiguous.

Some research topics are inherently 'sensitive' (i.e. psychological trauma, loss,
bereavement, discrimination, sexism, or suicide). Merely discussing these subjects
can be perceived as painful. Similarly, some techniques necessitate a physical
response from participants and can result in some harm. Does that imply these subjects cannot be researched, and that these stimuli cannot be used? Not necessarily.

In an experiment that provides a telling example, researchers tried to establish the causal relationship between workload and stress response. To do so, they had to induce a potentially harmful stimulus, namely some form of stress. The results showed that such stimuli do indeed have an influence on a participant's perceived well-being and impacted their physical health, as indicated by an increased cardiovascular response (see Hjortskov et al. 2004).

Is it justifiable to expose respondents to harmful stimuli, even when the effect is likely short-term? Hjortskov et al. answered the question in the affirmative and took refuge in what is considered by many as a safe baseline in research ethics. If harm does not exceed *the equivalence of what can be expected to occur in everyday life*, they argued, then the procedure should be safe.

It has been maintained that invasive techniques using stressors, unpleasant noises, rude or unkind remarks, among other forms of aggravators, are acceptable when (a) there are no other non-invasive techniques at hand, (b) the effects are equivalent to what people can expect to encounter in everyday situations, (c) have no long-lasting impact, and (d) everything is done to minimize harm.

Some retort that this will not (always) be sufficient. People who face systematic stigmatization in everyday life, or social exclusion, would be harmed in a way that is not acceptable should they be exposed to such stimuli, even though that is exactly what they expect to occur in everyday life.

Q4: Will the research design procedures result in any (unacceptable form of) harm or risk?

- Which possible risks of harm are feasible in this research?
- How do you plan to minimize harm (if any)?

10.6 Deception

10.6.1 Deceptive Techniques

Any research procedure in which a participant is deliberately provided with misinformation is labeled as a *deceptive technique*. Deception involves (a) giving false information, or (b) generating false assumptions, or (c) withholding any information that participants may request, or (d) withholding information that is relevant to appropriate informed consent (Lawson 2001, p. 120). Just because early research on the harmfulness of deception does not indicate that deceived participants feel harmed (Christensen 1988) or that they become resentful (Kimmel 1998), does not mean it is without moral consequence.

By default, deception excludes consent (see below). Participants are therefore not at liberty to decide to participate (or to continue participating) on the conditions known to them, regardless of whether consent was given afterwards, or even whether participants agreed to be deceived beforehand (when they agree to be fooled in some way).

Deception thus suggests a possible breach of two important ethical principles: the protection of people's autonomy and dignity, and the fair and equitable treatment of participants. Some have called for the abandonment of deception in research altogether, while others maintain that certain research areas, particularly in psychology, cannot do without it (see Christensen 1988). At any rate, IRBs have become more cautious in the last decade and generally insist on a full debriefing at minimum (see Mertens and Ginsberg 2009, p. 331). But even a full debriefing may not always be possible.

To summarize: forms of deception include providing false or misleading information about:

• Research goals or aims
• Research setup
• The researcher's identity
• The nature of a participants' tasks or role
• Any possible risks or consequences of participation.

The distinction between *false* information and *defective* information is noteworthy. False information means presenting an (oftentimes completely) wrong picture of the true research goals, while defective or misleading information might only mean withholding some (key) aspects thereof. Some argue that not telling participants certain things is not a form of deception (Hey 1998, p. 397), but we concur with Lawson (2001) that it certainly *can* be, especially on a relational level (pertaining to the relationship between researcher and participant) (Fig. 10.2).

10.6.2 Four Cases

Consider the following four cases in which (some form) of deception was deployed. How does the form and level of deception differ in these cases?

In the first, that came to the attention of one of the authors of this chapter, a group of researchers proposed to approach a number of intermediaries with mock job application letters and matching CV's that differed only with respect to the ethnicity of the 'applicant'. The researchers intended to measure the response rate of the

		Information Provided to Participant		
		Full	Incomplete/defective	None/False
Type of Consent Offered	Full consent	No deception (default)	Deception	Deception (questionable)
	Adhoc consent	Debriefing	Deception/debriefing	Deception (questionable)
	No consent	Tacit consent (questionable)	Objectionable / illicit deception	Objectionable / illicit deception

Fig. 10.2 Degrees of deception as a function of consent/debriefing and provision of information

intermediaries as an indication of hidden discrimination. The 'participants' (the intermediaries) were neither informed of nor debriefed about the research project, and thus would not be able to not participate or retort to its findings. Deception was deemed necessary to elicit true behavioral response.

The second pertained to an unpublished ethnographic study into social exclusion of the poor in Poland, carried out by one of the authors of this chapter. The researcher asked participants if they could be interviewed about their 'lifestyles,' deliberately not mentioning the goal of the study (social exclusion) because the researcher reasoned it might instill them (against their own conviction) with an idea that they are marginalized and excluded. The researcher feared that this idea would impose on them an identity that they could perceive as harmful. The research participants who were asked for consent were not informed about the true nature of the research project, nor were they debriefed afterwards. In this case, deception was considered both necessary as well as in the interest of the participants.

The third case concerns a covert participant observation project in an online anorexia support community performed by Brotsky and Giles (2007). The researchers created a mock identity of an anorexic young woman who said she wanted to continue losing weight. The researchers wanted to study the psychological support offered to her by the community, who was not informed about the research project. Throughout the course of the project, the invented character of the researchers developed close (online) relationships with some of its members. They justified the use of a manufactured identity on the grounds that if the purpose of the study was disclosed, access to the site would probably not be granted. Deception was deemed acceptable because of the 'potential benefit of our findings to the eating disorders clinical field' (2007, p. 96). The research participants were never asked for their consent, nor informed about the nature of the research.

The fourth case concerns social psychology research into the *bystander effect* (the inclination not to intervene in a situation when other people are present). Experiments on the bystander effect rely heavily on giving false information about the roles of other participants involved in the study, because they are in reality in cahoots with the researcher.

In a recent study into the bystander effect, Van Bommel et al. (2014) wanted to know whether the presence of security cameras would have any influence on said effect. The researchers designed a realistic face-to-face situation featuring a security camera (not featured in the control group). They exposed participants to a mock 'criminal act' to see whether they would respond or not. Immediately afterwards, participants were informed of the true nature of the setup.

In all cases, some form of deception was considered necessary, though for different reasons. Deception contributes to inequity between the research and the participant. By *debriefing* the participant (i.e. informing them of the true nature or purpose of the research), some of this can be countered under certain circumstances. In the first case discussed above, debriefing was not considered, in the second it was ruled out. In the fourth case it was part of the design by default and not questioned as such. In the third case, it *could* (and some would argue should) have been used.

Box 10.6: Checkbox for Ethical Concerns in Social Sciences Research Design

	Invasive/ intrusive	Deception/false information	Exposure to risk/harm	Vulnerable participants
Biometrics				
Randomized control trial (RCT)				
Laboratory experiment				
Experimental intervention				
Participant observation				
Survey				
Online survey				
Interview				
Vignette				

Which research techniques do you use in your design, and to what extent is your design vulnerable to the ethical concerns above? Provide a detailed description.

10.6.3   Deception and Misinformation

Arguably what matters most in considering the use of deception are found within two parameters: the degree of misinformation and the degree to which participants may give consent or can be debriefed (Ortmann and Hertwig 2002) (see Fig. 10.3). The questions any researcher must answer regardless are (1) whether or not it is really necessary to use deception, and (2) how to repair inequity if it were to be used.

Q5: Will the research design provide a full disclosure of all information relevant to the participant? If not, why not?

• How do you ensure your participants are adequately informed?
• What do you do to prevent deception?

10.7   Informed Consent

10.7.1   Informed Consent Protocols

Following established *informed consent protocols* are indispensable in any scientific research and serve to ensure that research is carried out in a manner that conforms to international regulations (such as the 1966 United Nations International Covenant on Civil and Political Rights, that explicitly prohibits that anyone be subjected to scientific experimentation without their permission).

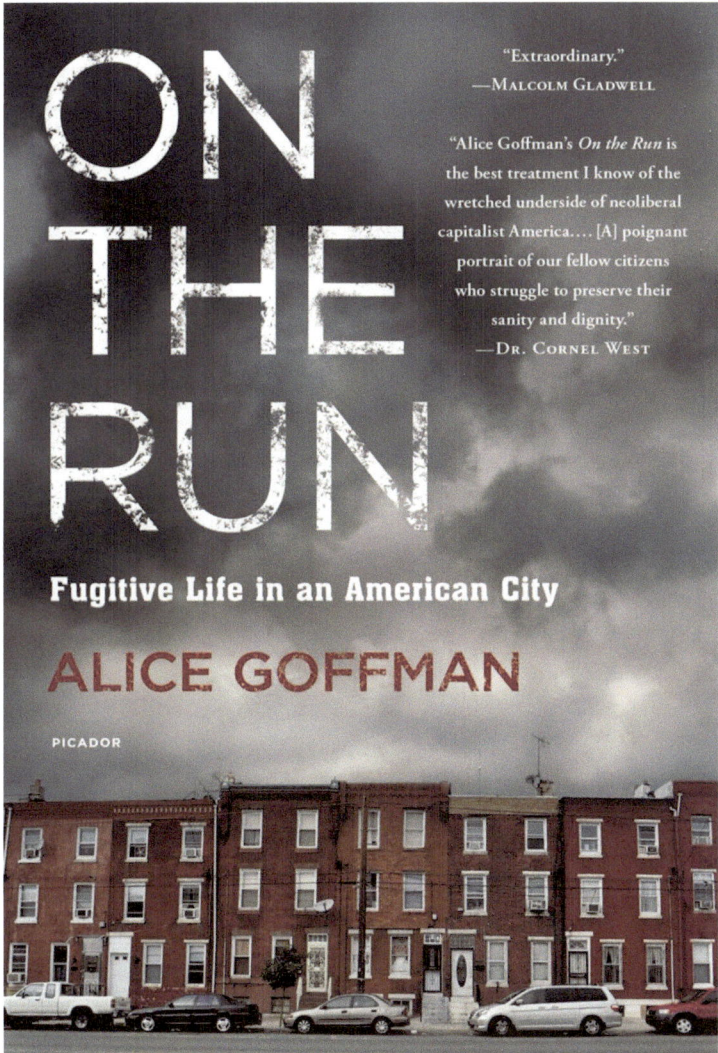

Fig. 10.3 Alice Goffman, *On the run*

Consent is based on four prerequisites: (1) it is given voluntarily (free from coercion), (2) the participant is a legally competent actor, (3) is well informed, and (4) comprehends what is asked of them.

To inform a participant means they must be notified about the objectives of the study, be informed about what is expected from them, and be told how their data will be used. Consent requires that a participant not only has a substantial understanding of the situation in which they will partake, but is also at liberty to refuse. Consent gives the researcher the right to involve the participant in the research project and at the same time assures that the respondent's rights are protected.

Informed consent protocols assume different shapes and forms. Traditionally they were hard copy forms, physically signed by the participant. Lately, they are often digital (i.e. in an online questionnaire, the respondent is informed about the objective of the questionnaire and needs to agree by ticking a box before proceeding). In ethnographic studies, speech recordings are sometimes utilized.

Consent protocols require furthermore that participants are provided with information of whom to turn to in case of disagreement complaints, or in case of unexpected or accidental findings that may affect the participant. This can be an independent board or a professional not involved in the study.

10.7.2   Who Can and Cannot Consent?

From our definition, it follows that any adult capable of understanding what is communicated to them and is at liberty to say no, can give consent. This leaves out a list of people who cannot be expected to give consent for reasons of incompetence, incomprehension, or lack of freedom. These include:

- Minors or children, who cannot legally give consent
- Adults with cognitive impairment or diminished decision-making abilities, who may not comprehend properly
- Adults in a dependent situation, such as refugees or undocumented immigrants, who may not be at liberty to refuse cooperation.

In some cases, others may consent for them (with children, this may be their parents or legal caregivers; with patients it can be their legal representatives). See Box 10.4 for an overview of vulnerability categories.

10.7.3   Active and Passive Consent

Consent is by default *active*, which means that the participant is knowledgeable about the purpose of the research and actively agrees to participate in it, under the conditions spelled out to them. *Passive consent* follows a different path. The participant is informed about the research, but it is assumed that they do not object, thus *passively* agreeing to participate. The researcher proceeds *unless* the participant actively refuses to participate.

Passive consent results in higher response rates and was more commonly sought after until the first decade of the twenty-first century. Today, a stricter view on subject autonomy is held, and consequently many IRBs no longer condone its practice, allowing it only in exceptional situations (Rangle et al. 2001).

10.7.4 Whose Responsibility?

It is the responsibility of the researcher that the participant fully understands what the research amounts to. The information provided must be comprehensive, to the point, and non-technical. Participants should be aware that they may refuse to (further) participate, withdraw their consent, and have their data removed from the study at any time (Box 10.7).

In some cases, it can be challenging to obtain informed consent, especially when participants are not accustomed with formal, written discourse, or come from a cultural background where such a formal permission would raise suspicion (see Israel 2015, for further discussion on the pitfalls that informed consent may carry, especially in qualitative research) (Box 10.10).

In other cases, consent may need to be re-negotiated. This can occur in longitudinal studies where parents earlier agreed that their children could participate in the study, but the child meanwhile grows up and becomes an adult capable of making their own decisions.

10.7.5 Disclosure of Sensitive Information in Consent

Researchers are obliged to conceal information that might be damaging to the respondent's reputation or affect their position within their community, organization, professional field, or could have an impact on their employability. For this reason, some institutions request that their researchers report only on data larger than a certain n-value, to prevent others from finding out who the participants may be (i.e. the Karolinksa Medical Institute at Stockholm set the norm at $n > 6$).

Such considerations are of relevance even when the study participants had, prior to their involvement in research, expressed their consent or even a wish to stay *non-anonymous*. The latter might happen with participants who are politicians or activists, who might treat their participation in research as a means to get publicity.

Q6: Have you obtained informed consent?

- What information have you communicated to your participants?
- In what ways have you ensured they are aware of what is expected from them?
- Check with your local IRB for samples of an informed consent and/or guidelines (see end of this chapter for a sample).

Box 10.7: Informed Consent as a Universal Principle?

Although *informed consent* is accepted world-wide as a necessary requirement for research, the question can nevertheless be posed whether or not it is
biased in favor of a Western notion of liberal individualism. Would the moral
conception and ideal of informed consent be applicable in China, whose cultural and ethical traditions are often conspicuously different from those of the
West, and are more specifically communally oriented (Nie 2001)?

Liang and Lu (2006) did research on legal reforms in China, for which they
conducted interviews. When seeking informed consent, they ran into what
they called a conflict between the *rigidity* and *inflexibility* of informed consent
and the *relativity* and *informality* of Chinese culture. For Chinese participants,
consent would be regarded not so much as a legal formality but rather the
foundation for continued friendship and trust. Consequently, they'd be hesitant to sign a consent form beforehand.

Furthermore, Chinese participants have a different view on the legal system. While Americans trust the confidentiality agreement because laws protect privileged information, in China no such laws or legal practices exist. As
a result, Liang and Lu wrote, 'a mere promise of confidentiality from the
researcher to the participant would indeed raise red flags about the legitimacy
of the research, therefore hurting rather than helping one's research' (2006,
p. 166).

Tangwa (2014) exemplifies the situation of West African women, who
because of bride-price practices, are in unequal and therefore vulnerable
relationships. These women are required by their communities to get
approval from their husbands if they volunteer to enlist in medical research;
by insisting only they themselves can give consent, their cause will not be
furthered.

Castellano (2014, p. 278) argues that the interests of Aboriginal
peoples are not served with individualized consent procedures. The implementation of ethical standards for Aboriginal research should be in the hand
of Aboriginal peoples. National committees should be formed, consisting of
Aboriginal experts, who could develop such standards, and help prevent misrepresentation and stereotyping, and ensure that environmental research is
included.

In these and similar cases, individualized informed consent procedures
are all but appropriate. Instead, consent extends to communities, experts,
or special committees, who oversee that interests of certain groups are
served.

10.8   Fieldwork and Data Collection

10.8.1   Entry Strategy and Conflicting Loyalties

When planning data collection, some considerations must be taken into account regarding strategies to access the 'field' (be that a school, a municipality, an internet community, or any another institute that houses participants) (Eysenbach and Till 2001).

Sometimes formal approval is required, other times approaching participants necessitates little more than the go-ahead from the head of an institute. Particularly when studying relatively small, tightly-knit communities or groups, caution is required. In those cases, researchers make use of *gatekeepers*, such as institutions or persons who have (direct) access to potential study participants. While gaining access to a field via gatekeepers has its obvious advantages, it may also involve some moral dilemmas.

Gatekeepers, just like research sponsors and research participants, usually have their own stake in research. When offering access to their networks, gatekeepers show trust and expect loyalty. Researchers thus engage in what is called *relational ethics*, which builds on mutual respect, dignity, and connectedness between the researcher and researched (Ellis 2014, p. 4), although the researcher often cannot avoid politically embedded issues of power that require a 'delicate balancing act' (McAreavey and Das 2013).

10.8.2   Cooperation and Non-Cooperation

Once people agree to participate in a study, the researcher may count on their cooperation and benevolence. At a minimum, participants should not feel deceived, intimidated, or otherwise uncomfortable with the research, but there can be many other valid reasons why participants decide to leave a study. In some cases, such as in evaluation studies that involve a researcher's prolonged presence in the field (perhaps even against the wish of some of the actual study participants), participants may become reluctant, mistrustful, or even non-cooperative. In other cases, participants my leave studies for no apparent reason at all (as they are free to do).

Research using large databases of raw, unstructured public data ('big data') poses its own ethical considerations, in particular with regard to privacy. In consumer behavior research, for example, Numan and Di Domenico (2012) observe that the volume and speed with which data must be analyzed often requires data collecting and analysis without an individual providing specific consent. This raises ethical concerns 'relating to the extent to which organizations can control the collection and analysis of data when there is limited human involvement' (2012, p. 51).

Finally, there is the question of non-cooperation (or counter-cooperation), which can occur in a variety of ways. Researchers may find that participants avoid answering certain questions, are purposefully manipulative, or even lie about particular issues (because of shame or to protect their dignity).

All these forms of non-cooperation will pose researchers with a challenge, and it is therefore advisable to think ahead to strategies for what to do in case there are not sufficient data points to work with.

10.8.3 Interpersonal Dynamics

Ethical dilemmas may also arise as a consequence of interpersonal dynamics, both between the researcher and the study participants and (or) among study participants. In any case study that involves participant observation or repeated interviews with participants, continued interaction is likely to result in emotional and social engagement on the part of the researcher. This may lead to the formation of alliances and conflicting loyalties. As a result, the role of the researcher as an 'objective observer' of social life might be challenged.

In the course of any study, the researcher's relationship with the gatekeepers or sponsors may also need to be renegotiated, for example, when gatekeepers try to influence the direction of the study. Commitments related to confidentiality and anonymity may need to be re-affirmed or redefined. Finally, in cases of intensive ethnographic observations, the prolonged presence of the researcher is also likely to re-define the community or group or organization studied, and may raise (moral) questions related to the role of the researcher and their relationship with the object of investigation (Mikesell et al. 2013) (Box 10.8).

Q7: How do you enter the 'field'?

- Which formal or informal agreements have you made and with whom?
- Which expectations have been created when an agreement on cooperation has been made?
- How do you deal with non-cooperation on the part of study participants?
- How do you deal with competing loyalties?

10.9 Incidental Findings

10.9.1 Incidental Findings in Clinical Research

Any research, including the most non-invasive varieties, can unearth *Incidental Findings* (IFs). For example, brain imaging research may bring to light undetected, clinically relevant abnormalities that are unexpectedly discovered and although unrelated to the purpose of the study, they may require urgent or immediate referral (Vernooij et al. 2007).

Box 10.8: Alice Goffman – What are the Limits of a Researchers' Involvement?
Sociologist Alice Goffman's 2014 ethnographic study *On the Run: Fugitive Life in an American City* details the careers of poor black men in West Philadelphia. She paints a bleak picture of these men who follow 'the other path into adulthood,' leading them invariably to crime and eventually incarceration.

In the wake of the 'Black Lives Matter' movement, the book was received with ample praise. Goffman's central claim, that it is the legal system itself creating crime and dysfunction in poor black communities, was supported by her critics, although some reviewers would argue that Goffman's views are rather one-sided. For example, Heather Douglas (2014) wrote that Goffman refused to acknowledge that her participants create their own predicament through deliberate involvement in crime.

From a research ethics viewpoint, *On the Run* raises an alarming question about the limits of a researchers' involvement. Some accounts in the book suggest that Goffman was so thoroughly involved with her participants that she became complicit in criminal activity herself, including even conspiring to commit murder (as one participant confided in his plans to kill someone). She thus violated perhaps the most basic precept of scholarly (and personal) responsibility – not to endanger somebody else's life, and to do no harm (Lubet 2015). Whether she committed any crimes cannot be established, as she had carefully concealed the true identity of the participants involved and destroyed her field notes, which from an ethical viewpoint is also questionable.

This raises the question of what to do with these findings. Is it in the interest of the participant that the researcher notifies them? The intuitive answer may be yes, but some argue that this is not evidently the case, as participants have a right *not to know*.

On the one hand, Miller et al. (2008) argue that clinical investigators do have an obligation to respond to incidental findings. They argue this point because the researcher entered into a professional relationship with the research participants, and thus they are granted privileged access to private information with potential relevance to the participants' health. Appelbaum et al. (1987) on the other hand, warn against the false hopes that a confusion of roles might create, when participants feel that research protocols are designed to benefit them directly rather than to test or compare treatment methods.

Incidental findings call for the weighing of *false positives* (potential harm due to findings that have no clinical significance) against *false negatives* (failure to report a finding linked to a serious health problem). In an attempt to solve at least part of this quandary, IRBs often suggest that participants be offered an 'opt-in' / 'opt-out' choice in the informed consent. 'Opt-in' necessitates the researcher to communicate any accidental findings relevant to the participant, 'opt-out' prohibits the same.

However, even with such clear-cut arrangements, the researcher may still find it ethically problematic to remain silent when the participant chooses to 'opt-out' and a clearly identified, life-threatening, treatable condition is discovered (for further discussion see Illes et al. 2006).

10.9.2 Incidental Findings in Non-Clinical Research

DNA analysis is increasingly utilized in forensic anthropology, for example to identify damaged or fragmented human remains, for which DNA of family members is required for proper identification. Parker et al. (2012) argue that the increased prevalence of incidental findings (IFs) in non-clinical research (such as misattributed paternity or false beliefs about sibling relationships), calls for new policies to focused on *minimizing* the discovery of IFs.

Q8: How should you deal with incidental findings?

• Which agreements have you made with your participants regarding any potential incidental findings?
• Do you offer an opt-in/opt-out option in your informed consent?
• How do you check for any unintended consequences of discovering incidental findings?

10.10   Analysis and Interpretation

10.10.1 Analyzing Results

While fraud among academics is rare, questionable research practices do occur, leading to multiple forms of bias. These include (among others): *publication bias* (non-publication of null results), *confirmation bias* (tendency to look for confirmative results, disregarding of contradictory information), and *funding bias* (tendency to support the study's financial supporter) (Box 10.9).

We believe that it is vitally important that researchers are aware of the forms of bias that lie in wait and operate as transparently as possible (see Chaps. 5 and 6 of this book for an extensive discussion). Below, we discuss three issues related to bias, namely *significance*, *plausible objection*, and *limits of interpretation*.

Box 10.9: Transparency: Steps Researchers Can Take to Reduce the Risk of Bias

Dr. Daniele Fanelli: 'The way out of [bias], generally speaking, is to be transparent about what you did. I'm not naive enough to think that this is going to be the whole story, because publication space in journals is limited, and you will never be allowed to tell precisely everything that you have done. So in part, the system does need other ways also to allow researchers to make fully public their data, you know, all the results they obtained, etc.

Again the ideal to follow, I think, in any kind of research, is as much as possible, be transparent of the whole procedure. What were your original research questions, how you collected the data, what eventually was the data that went into this particular study, and so on.

(From online course site Epigeum, Research integrity: arts and humanities, module 3).

10.10.2 Significance

After having collected data, the main concern of a researcher is whether or not the data has a story to tell. Are the results indeed significant? Is the effect a sizeable proportion? Is there a convincing narrative pattern discernable in the interviews?

Given the study produced at least *some* results worthy of discussion, then the ethical question is: are the results significant and unique enough to warrant publication? This is not self-evident nor easily established. What is 'significant' in this context does not depend on a statistical or discursive measure, but on an overall evaluation of the results. This evaluation would also include the question of whether the phenomena observed was properly accounted for by available theories. Finding (statistically) significant results is one thing, finding something that is substantial is something else. It is the task of the researcher to carefully assess the weight of their findings.

10.10.3 Plausible Objection

Complementary to the question of significance is the question of *plausible objection*. One must ask themselves, if a study produces data containing significant results, how do those results line up with rivalling theories or other plausible explanations and objections? How do we know that novel findings were truly revealed, and not merely an exception to the rule, chance findings, or even false positives?

Again, the answer to these questions cannot be established on the basis of the data alone, but they need full consideration, nonetheless. The place to address these considerations is often in the discussion section of a scientific paper or as an addendum to the findings, although in reality they should precede any discussion of the findings.

10.10.4   Limits of Interpretation

Findings worthy of publication need to be framed such that their significance can be understood and eventually be communicated to others. But how far can our interpretation go?

In 2002, the then United States Secretary of Defense, Donald Rumsfeld, conceived of a way to interpret uncertain knowledge. He invented a concept that since found its way into scientific parlor: the 'unknown unknowns.' It plays a role in a distinction between 'known knowns' (the results of a study; certain evidence) and 'known unknowns' (certain variables not researched; certain contexts not taken into account). 'Unknown unknowns' are possible risks, future outcomes, or consequences one is not aware of.

While 'known unknowns' may point to the direction of future research, 'unknown unknowns' point to the fundamental uncertainty in scientific research. By *imagining* possible events and occurrences, certain 'unknown unknowns' may become 'known unknowns.' Similarly, by pointing out certain 'blind spots' in a frame of reference, 'unknown unknowns' may become 'known unknowns,' or even 'known knowns' once they are researched (Box 10.10).

Q9: What is the significance of your data?

- What do your findings say in respect to alternative explanations and plausible objections?
- What are the limits of interpretation of your study?

10.11   Reporting and Dissemination of Research Findings

10.11.1   Dissemination and Responsibility

Research findings are commonly reported in scientific outlets, such as peer reviewed journals and scholarly books, or at international congresses and academic conferences. Alternative ways to reach various other audiences may include: articles in popular journals and newspapers; brochures and leaflets tageting certain lay audiences; appearances in seminars for professionals; participation in think tank research; radio and television performances; hosting of podcasts; involvement in internet forums; providing training sessions; and individual or group counseling (for further discussion see Oliver 2010).

Box 10.10: Qualitative Research
In qualitative research, many of the steps described in this chapter cannot be separated clearly, but rather merge under certain conditions. The following flowchart, borrowed from Damianakis and Woodford (2012, p. 715) details various considerations when planning a research project in small connected communities, recruiting participants, collecting data, and disseminating results (Fig. 10.4).

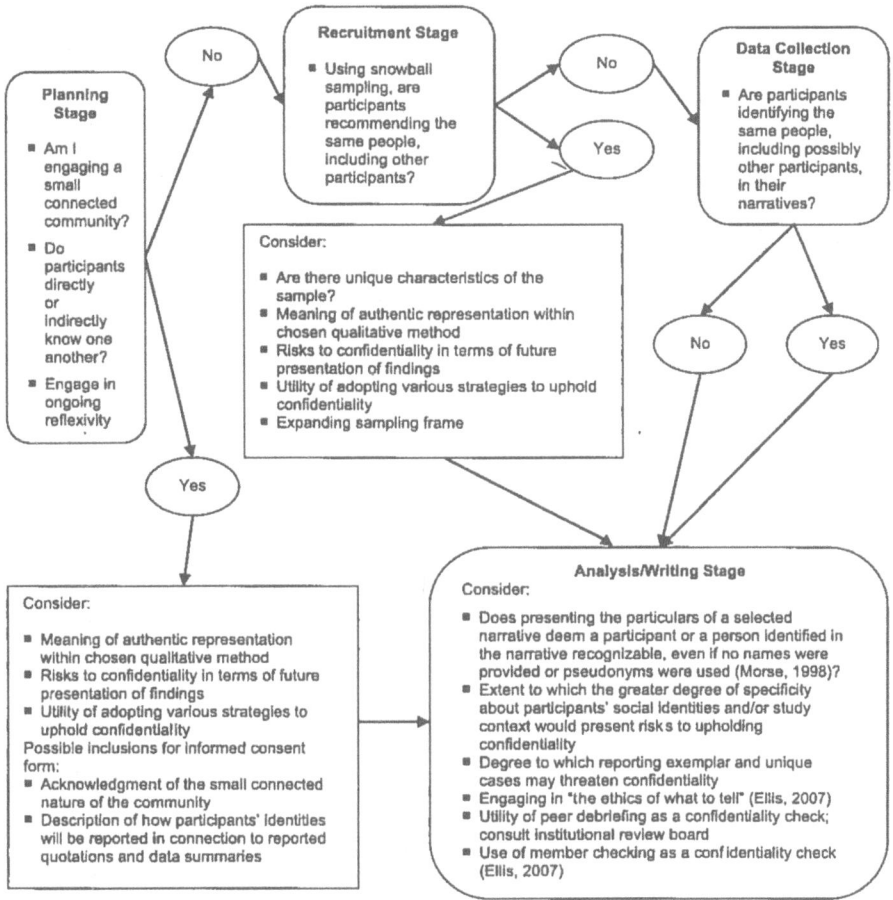

Fig. 10.4 Considerations when planning a research project (Damianakis and Woodford 2012, p. 715)

We shall not discuss all of these individual forms of communicating research findings, but instead we will flesh out the ethical implications of three different *role responsibilities* (Mitcham 2003) inherent in the task of a researcher engaging with their audiences.

10.11.2 Responsibility to Participant

Researchers carry a responsibility to inform their participants about the results of their work. How do they fulfill this requirement? Often they suffice to just offer the possibility to learn about the findings of their research projects, and this usually entails no more than notifying the participants when the report is published.

In many cases, this may be sufficient, especially when participants have simply filled out a questionnaire or took part in an experiment, and were not otherwise involved in the research project. Considering that oftentimes research participants are not typical readers of academic journals, no further action may be required on the part of the researcher.

However, if, as in qualitative research, or in action research, respondents have dedicated time and energy into research projects, or are involved in it to some degree, the responsibility of the researcher to inform them of their research findings would not end there.

In either case it may be worthwhile to consider how respondents are affected by the research, and whether or not some 'aftercare' is needed in the form of ad hoc reevaluation or debriefing (Box 10.11).

10.11.3 Responsibility to the Research Community

There is also an obligation to communicate research findings to the scientific community, and this obligation goes hand in hand with the requirement to be critical of one another's work in service of furthering scientific knowledge. There are a few issues that can be raised here too.

Academic engagement with private industry is rapidly growing, and this is impacting academic research, as a literature review by Perkman et al. (2013) reveals. Commercialization may enhance productivity (on the short run), but it also impacts the agendas of researchers and promotes an environment of secrecy.

Although research commissioned by third parties is becoming more prevalent, it should be made perfectly clear whose interests are at stake. It has therefore become common practice for researchers to be required to disclose any affiliations with outside institutions; reveal specific financial arrangements, including arrangements concerning intellectual property; as well as divulge any other ties of a social, political, or personal nature that might indicate a conflict of interest. In short, researchers must be hyper-transparent (for further discussion, see Chap. 8).

Box 10.11: Whose Perspective Prevails?
Reporting research findings can be precarious, as the following example of an unpublished qualitative study reveals. The research project, financially supported by several municipalities, aimed to analyze the perspectives of policy makers, healthcare providers, and their clients on homeless shelters.

In several interviews, some of the clients (homeless individuals) complained about the poor living conditions in one of the shelters, and the inadequate support offered by one of the healthcare providing parties therein. One interviewee said: 'It's a mess. At least that's how I see it. If you want to help people, you should do it completely different.' Several of these complaints were included in the first report released, which was subsequently sent to the parties involved in the research project, including the healthcare providers.

Shortly thereafter, an argument arose between the researcher and the healthcare provider that had been criticized. The healthcare provider objected to the 'uncritical publication' of these complaints, which they believed were baseless and even harmful.

The researcher offered the healthcare provider an opportunity to contradict the complaints in a separate section, which would be inserted as an addendum in the next report sent to the relevant parties. The healthcare provider declined, insisting that it should be the researcher *themselves* that rectified what they considered to be a 'grave mistake.' The researcher's supposed portrayal of a 'crooked image' of the organization would cause them serious damage, the healthcare provider argued.

The researcher objected, contending that it was their academic duty to report *all* research findings and not to favor one party. The healthcare provider thereupon threatened to file an official complaint against the researcher, with whom they would no longer collaborate if the researcher would not concede.

How would you advise this researcher? Should they:

1. Back down, revise the text, and omit the complaints to rescue the working relationship with the healthcare provider?
2. Persevere as a scientist who has the duty to report findings as objectively as possible, even at the cost of a working relationship?

Which option would you choose, and why?

(Case was communicated to one of the authors of this chapter.)

10.11.4   Responsibility to Civic and Professional Communities

Disseminating knowledge to civic and professional communities entails a different focus than those of academic communities. Inasmuch as sharing knowledge is geared towards the application of theoretical insights, the information provided needs to be of practical value, which can be used for the purpose of training, evaluation, risk assessment, or other needs (Fig. 10.5).

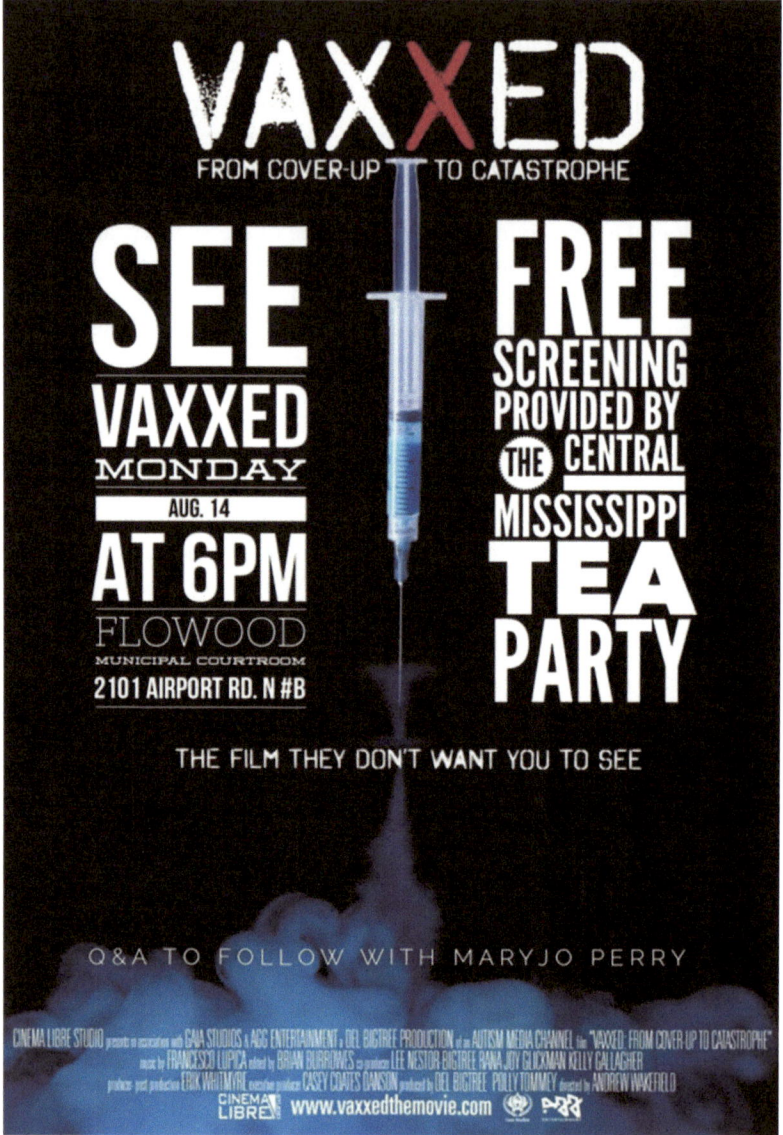

Fig. 10.5 'Vaxxed', the 2016 documentary defending Wakefield and his followers

This means a shift away from asking whether research conclusions are true (the focus of academic knowledge dissemination), toward asking under what conditions and assumptions the findings are valid (the focus of civic and professional knowledge dissemination), and this task has a number of ethical implications. This is especially true when vulnerable communities or developing countries are involved and questions of social responsibility emerges. What are the researchers' obligations towards these communities? Due to their economic, political, and intellectual power, what are the duties of the scientist in relation to society and public interests (Payne, 2000) (Box 10.12)?

Q10: Which information are you obliged to share with your research participants?

- Which ties, commitments, and affiliations are of importance to the understanding of your research?
- Which moral responsibilities do you have towards others, specifically to those with a stake in your research?

Box 10.12: The Anti-Vax Movement. Who is Responsible for What?

A 1998 paper published in *The Lancet* (a high-ranking peer reviewed medical journal) connecting MMR (Measles, Mumps and Rubella) vaccination and the onset of autism sparked a worldwide anti-vaccination movement, raging still today.

The paper describes twelve children, most of whom were diagnosed with autism, who had bowel and behavioral problems. Eight of those children were purported to have developed autistic symptoms within a few days after they had been given the MMR vaccine. The story was picked up by several large UK newspapers and became the center of a nationwide debate in the subsequent years. This fervor eventually spread to other countries, which brought the paper's principal author, Andrew Wakefield, considerable fame but also substantial criticism.

The scope of the paper was very limited (it was merely a series of 'case reports' and the sample size was only twelve). Based upon scientific norms, it would not have led to any reputable conclusion about a relationship between the MMR vaccination and autism. In fact, subsequent studies failed to find *any* such connection. These later studies, that did systematically probe the relationship between MMR vaccines and autism, received far less media coverage than the original Wakefield paper (see Ben Goldacre's 2009 *Bad Science* for details about the study and its media reception).

The Lancet study was retracted when it was found to be seriously flawed on several accounts. Data had been falsified and the research was deemed unethical because of Wakefield's 'callous disregard for any distress or pain the children might suffer.' Additionally, it was argued that the author himself was compromised because of undisclosed financial conflicts of interest. Wakefield's medical license was revoked in 2010.

All of this did nothing to deter the anti-vax movement (of which Wakefield was, and still is, the poster child) or halt its momentum. There was even a 2016 documentary with a pro-Wakefield spin, called *Vaxxed,* which in the tradition of conspiracy theories, was advertised: 'from cover-up to catastrophe.'

The number of people who refuse to vaccine their children has now risen to dangerous levels. These individuals believe, misinformed as they may be, that measles is harmless, vaccines are dangerous, and that the government has

---

(continued)

Box 10.12 (continued)

no business interfering in their lives. Even worse, some believe that the government conceals 'the truth' for the sake of a 'powerful medical-industrial complex.' The result of this flood of misinformation? Rates of this deadly disease have begun increasing yet again.

The anti-vax case raises the serious question of who is to blame and for what are they to be blamed. Framed differently, where does a researcher's responsibility begin and where does it end? And when do outside parties begin to share in this responsibility?

How would you define the responsibilities of the following actors in this anti-vax case, with regard to their obligation to communicate scientific findings?

Start with Wakefield, as the Principal Investigator (PI) of the study, who has an obligation to report not only truthfully but also responsibly about his research findings. Given that he honestly believes that MMR vaccines relate to (or even cause the onset of) autism, what responsibilities do you think he has as a scientist to communicate his findings? Is he to be blamed for what some consider a dangerous hoax? And how about the other parties involved in this case? Flesh out the responsibilities of all the parties involved as best as you can.

Actor	Responsibility
Wakefield (as PI of the study)	
Editor of *The Lancet*	
Editors and journalists of newspapers	
Documentary filmmakers	
Medical researchers	
Governments / authorities	
Medical doctors	
Anti-vaxxers	

10.12   Data Management and Storage

10.12.1   Secure Storage

Secure storage of research data is at the core of research ethics, especially today, in the age of hacking and data breaches, and is subject to expanding regulation worldwide. It serves two basic purposes: *verification* and *reuse* (for secondary analysis).

At first sight, it may appear desirable to simply preserve all data collected during research and to make it available to any and all researchers, in order to prevent fraud and render science more efficient. However, the issue in preserving research data touches on its confidential nature (see Chap. 7). Additionally, the competing interests of researchers, research participants, and other stakeholders presents a number of challenges. To deepen the challenge, conflicting database legislation in different countries makes the preservation of all research data near impossible.

Decisions have to be made as to whether or not data will be made available, to *whom* will have access, *how* it will be accessed, *where* it will be stored, and *how long* it will be stored (for an overview of these considerations, see Johnson and Bullock 2009). We will briefly discuss the main aspects of these questions below.

10.12.2 Sensitive Data

Before decisions are made about whether data should be archived and shared, the data's sensitivity should be assessed first. What is deemed 'sensitive' in a legal sense may differ from country to country, though many would agree that any data containing personal information would classify as such. Sensitive data would thus include a participant's identity, information about their ethnicity, gender, political opinions, medical history, sexual orientation, religious background, and philosophical beliefs.

How should researcher's deal with sensitive data? Several strategies have been developed to confront this important issue.

Anonymization Data is stripped of its identifying properties by assigning a code to specific pieces of information. For example, the name of the participant is replaced by a number. If a key is preserved that enables re-identification (linking names to number, for example), privacy policy requires that the key be stored separately, and shall not be shared.

Other policies insist, however, that no key be kept at all, and that data collection be anonymized right from the beginning, such that all data effectively becomes anonymous the moment it is collected, and can never be linked to individual participants. This strategy is most fitting for quantitative research practices.

Pseudonymization The true identity of the research participant or interviewee is concealed by giving them a pseudonym (an alias) and by changing other identifying details that might make identification possible. This strategy is more commonly practiced in qualitative studies, such as ethnography and case histories.

Some regard pseudonymization as an insufficient guarantee of privacy, as clever detective work may enable the identification of participants. For example, almost all of Freud's patients have since been identified, even though he went to great lengths to hide their identities. Destroying field notes to protect a respondents' privacy might seem to be a solution to this issue, though in practice, it raises questions of its own.

Co-ownership In some research practices, respondents define the goals of the research project in close collaboration with the researchers and remain actively involved in other stages of the project, including the interpretation and dissemination of the results. Typically, this is the case in 'action research' or 'community engaged research' (see Friedman Ross et al. 2009).

By becoming closely involved, research participants become co-owners of the research project, but this often means that the researcher cannot offer the same ethical guarantees concerning confidentiality and anonymity, informed consent, and protection from harm as in other research methodologies (Williamson and Prosser 2002). For example, when school professionals conduct action research, confidentiality will be much more difficult to secure (Nolen and van der Putten 2007).

10.12.3   Making Data available to Whom?

There appears to be a near consensus that data should be archived (if only for reasons of verification). There is dissent, however, over whether secondary researchers or other parties should be allowed access to said archived research data, even after anonymization or pseudonymization.

Large longitudinal research projects almost by definition require the sharing of data, if only for reasons of efficiency. However, legislation in many countries has become much more stringent about protecting the rights of research participants, and this can become an obstacle in these projects.

Legislation safeguards the rights of participants to:

Have access to their own data
Have their data corrected or removed at their request
Refuse any other use of their data than agreed upon

Should the foundational principles of privacy be followed strictly, as some argue, no other researchers should be allowed access to data *unless* participants consent to secondary analysis. Others, however, maintain a more liberal perspective, arguing that if data is entirely anonymized, then these restricting conditions need not apply. But even if that is the case, collaboration between teams of researchers from different countries can become quite difficult given that each country may possess different privacy rules.

10.12.4   Storing Data Where?

Securing data implies storage in a safe place. This could be an encrypted university hard drive, or the implementation of encryption software. Agreements must be made in advance as to who has access, and to which parts of the hard drive. Additionally, the question is who will maintain the data once it is stored.

For security reasons, data should never be kept on personal computers, laptops, or other information carriers. Similarly, hardcopy receptacles of sensitive information should be kept in safe places, such as a vault or a locker that can opened by designated people only.

Finally, some considerations must be given to possible data breaches, data leakages, and the accidental loss of sensitive data. What are the procedures that must be followed in the event that sensitive information is lost, or even made public by accident? Who needs to be informed, and who has which responsibilities?

10.12.5   Archiving Data for How Long?

Lastly, decisions must be made as to *how long* data should be stored. Some conflicts of interest may arise here. Some universities and research institutions insist on the extended storage of data (at least 10 years) for verification purposes, to prevent fraud and/or uncover forms of misconduct. This requirement, however, conflicts with certain privacy laws, which may demand the destruction of unnecessary data as soon as possible. It may also conflict with contractual obligations made with study participants (for example, when consent to participate in a study is given under the condition that collected data is destroyed immediately after the study report has been published).

Q11: How should you ensure that any sensitive data is rendered in a form that is fitting for the research purpose and stored in a safe manner?

- With whom can your data safely be shared?
- What are the data security and safe storage procedures at your institute or university? How do they differ from the agreements you've made with you research participants?
- What is a safe amount of time to archive sensitive data?

10.13   Conclusions

10.13.1   Summary

In this chapter, we followed a step by step approach to the ethical questions you need to answer when planning a research project. The objective was to learn what questions to ask, and to reflect on the answers as you plan and design a research project (see Box 10.6).

First, we discussed what research questions must be asked, to consider how important they are, and to think about what your research can contribute to. This was followed by a cost-benefit analysis of the risks inherent to research in the social sciences.

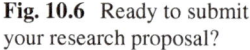

Fig. 10.6 Ready to submit
your research proposal?

Second, we examined the various implications in using a variety of research techniques, including the *invasion of integrity* and the risks of *deception*. We followed this with a brief outline of *informed consent protocols* and how you can avoid harm and do good.

Third, we considered the ethical issues involved in *collecting*, *analyzing*, and *interpreting* data. What (un)intended consequences does your presence in the field have on the participants and research outcomes? What promises do you have to live up to?

Finally, we reviewed the responsibilities that come with being a researcher, specifically when sharing your findings with others. We concluded with a discussion of the various issues involved in safely storing data.

Followed from start to finish, this chapter aimed to ensure social science researchers were made aware of any potential ethical pitfalls that may be encountered. Are you ready now to submit your research proposal? (Fig. 10.6)

Suggested Reading

We highly recommend *Research Ethics and Integrity for Social Scientists* (Israel 2015) and *The Student's Guide of Research Ethics* (Oliver 2010) as excellent reference books for students who want to learn more about the principles and philosophies of research ethics. Israel in particular gives an overview of the varying ethical policies found throughout the world. *The Handbook of Social Research Ethics* (edited by Mertens and Ginsberg 2009) offers an excellent selection of essays on a wide variety of topics in the history, theory, philosophy, and implementation of applied social research ethics. Especially worth mentioning is Chap. 8, on IRBs, written by Spiegelman and Spears. *Principles of Biomedical Ethics* (Beauchamp and Childress 2001) is a classic in the field of research ethics. Diener and Crandall's *Ethics in Social and Behavioral Research* (1978) offers an older yet still relevant insight into the field of research ethics as it first emerged.

10.14   Ethics Checklist

The following checklist may be useful when designing a research project. It is designed for students who do research under the supervision of a qualified researcher and can be adjusted at will and according to their own needs. It is emphatically not meant to replace local IRB's protocols.

Project details	
Project title	
Applicant details	
Name of student(s)	
Email address / student id number	
University / department	
Course name (if applicable)	
Supervisor's name	
Duration (from / to)	
Research project	
Please provide a brief outline of your study. What is its **purpose**? What are the main **theoretical assumptions**? What is/are the **research question(s)**? (ca. 200–300 words) Outline: Research questions: Hypotheses:	
Please answer the following questions to the best of your knowledge. Consult supervisor if needed.	
1. **Participants**	
What is the (estimated) number of participants. What is the power analysis to determine sample size, if relevant?	
Does the study involve participants who are unable to give informed consent (i.e. people with learning disabilities)? If yes: Explain why and what measures you will take to avoid or minimize harm.	
Does the research involve potentially vulnerable groups (i.e. children, people with cognitive impairment, or those in dependent relationships)? If yes: Explain why and what measures you will take to avoid or minimize harm.	
Will the study require the cooperation of a gatekeeper for initial access to the groups or individuals to be recruited? (i.e. students at school, members of self- help group, residents of nursing home)? If yes: Who is the gatekeeper? What agreement have you made, and which expectations do you share?	
Will it be necessary for participants to take part in the study without their knowledge and consent at the time (i.e. covert observation of people in non-public places)? If yes: Explain why and how, and provide a risk analysis if applicable.	
Will any dependent relationships exist between anyone involved in the recruitment pool of potential participants? If yes: Explain why and how, and provide a risk analysis	
2. **Research design and data collection**	
Will the study involve the discussion of sensitive topics? (i.e. sexual activity, drug use, politics) if yes: Which topics will be discussed or investigated, and what risk is involved? What measures have you taken to minimize any risk, if applicable?	

Are drugs, placebos, or other substances (i.e. food substances, vitamins) to be administered to the study participants? If yes: Explain the procedure and provide a brief cost-benefit analysis. What measures have you taken to minimize any risk, if applicable?

Will the study involve invasive, intrusive, or potentially harmful procedures of any kind? If yes: Explain the procedure and provide a brief cost-benefit analysis. What measures have you taken to minimize any risk, if applicable?

Could the study induce psychological stress, discomfort, anxiety, cause harm, or have negative consequences beyond the risks encountered in everyday life? If yes: Clarify the procedure and explain why no alternative method could be used. Provide a brief cost-benefit analysis if necessary. What measures have you taken to minimize any risk, if applicable?

Will the study involve prolonged or repetitive testing? If yes: Explain the procedure and clarify how the interests of the participants are safeguarded.

Is there any form of deception (misinformation about the goal of the study) involved? If yes: Explain the procedure and provide a rationale for its use.

Will you be using methods that allow visual and/or vocal identification of respondents? If so: What will you do to guarantee anonymity and confidentiality?

Will you be collecting information through a third party? If yes: Who is that party? Provide a brief outline of the procedure.

Will the research involve respondents on the internet? If yes: How do you plan to anonymize the participants?

How will you guarantee anonymity and confidentiality? Outline your procedure and give an estimate of the risk of a breach of confidentiality.

What information in the informed consent will participants be given about the research? Provide a brief summary or upload the consent form. Which procedures are in place in case participants which to file a complaint?

Will financial compensation will be offered to participants? Provide a short accounting of any compensation being offered.

If your research changes, how will consent be renegotiated?

3. **Analysis and interpretation**

What is the expected outcome of your research? What would you consider a significant result?

During the course of research, how will unforeseen or adverse events be managed (i.e., do you have procedures in place to deal with concerning disclosures from vulnerable participants)?

4. **Dissemination**

How do you plan to share your research findings? Which audience to you intend to target?

5. **Data storage**

Where will your data be stored? Which measures have you taken to make sure it is secure?

Which safety precautions have you arranged for in case of data leakage?

Will your data be disposed of? If yes: When? (date) if no: Why not?

Will your research involve the sharing of data or confidential information beyond the initial consent given (such as with other parties)? What specific arrangement have you made and with whom?

Principal investigator / teacher

Signed:	Date:	Place:

Student

Signed:	Date:	Place:

10.15 Sample Informed Consent Form

[Adapt this form to your proposed research project].

Information about Participation in a Research Study at [your university or research institute]

[Title of the study:]

INTRODUCTION: Thank you for taking part in this study about [give brief explanation of the study]. Below is a description of the research procedures and an explanation of your rights as a research participant. In accordance with the ethics code of the [local institution], you are asked to read this information carefully. You are entitled to receive a copy of this form should you agree to proceed under the terms stated.

GENERAL INFORMATION: The purpose of this research is [give brief description of study purpose here]. This research is funded by [insert here, if applcable]. The potential conflicts of interest are [describe any that are known]. [OR] There are no known conflicts of interest in the conducting of this research study.

Your participation will last for approximately [duration estimate] and will take place at [location] at the following times [dates/times]. You will receive [insert reimbursement, i.e.. number of PPU, amount of money, a chance to win a voucher or, 'no reimbursement'] for your participation in this study.

PROCEDURE: During this study, you will be asked to [insert brief description of what the participant will do]. You are aware that [describe any risks that are known]. [OR] There are no known or anticipated risks associated with participation in this study.

You have the right to end your participation at any moment, without citing a reason. If you choose to end your participation before the study terminates, you [will / will not] be reimbursed.

Regarding the use of your data, the following conditions apply:

- Your data will be used for scientific purposes, including publication. Only the researchers have access to the data [OR] The data will be made available for other researchers on condition of confidentiality.
- Your data will be handled and stored confidentially. This means that your data cannot be traced back to you. Specifically, the researcher will use a code number instead of your name to save your data.
- [If the data from the study will be personally identifiable] The code number and other personally identifiable information, such as names, will be saved separately from each other in a secure location.
- After publication, only the data that is necessary for the verification of the study results will be kept and stored safely for a minimum of 10 years and deleted once it is no longer needed. [OR] Personally identifiable data will be shared only if it is scientifically required to verify reported results.
- You have the right to withhold any responses you have provided from subsequent analysis. This means we will not use your data for this or any follow-up research, nor will we share it anonymously for open science purposes. You can decide to

withdraw your data until the study results are accepted for publication, or until the data is cleared of any and all identifying information, such that no-one will be able to trace you.

OFFER TO ANSWER QUESTIONS: You are now given the opportunity to ask questions. If you have any further questions or complaints about this study, you may contact the researcher, [name(s) of researcher(s) and email(s)—phone number(s) can be added if researchers prefer to use that method], of [your university or research institute].

References

Appelbaum, P. S., Roth, L. H., Lidz, C. W., Benson, P., & Winslade, W. (1987). False hopes and best data: Consent to research and the therapeutic misconception. *Hastings Center Report, 17*(2), 20–24. https://doi.org/10.2307/3562038.

Anderson, E. E. (2019). A proposal for fair compensation for research participants. *The American Journal of Bioethics, 19*(9), 62–64. https://doi.org/10.1080/15265161.2019.1630501.

Baumrind, D. (1964). Some thoughts of ethics of research: After reading Milgram's study of obedience. *American Psychologist, 19*(6), 421–423. https://psycnet.apa.org/doi/10.1037/h0040128.

Beauchamp, T. L., & Childress, J. F. (2001). *Principles of biomedical ethics*. Oxford: Oxford University Press.

Beck, H. P., Levinson, S., & Irons, G. (2009). Finding little Albert: A journey to John B. Watson's infant laboratory. *American Psychologist, 64*(7), 605–614. https://psycnet.apa.org/doi/10.1037/a0017234.

Becker-Blease, K. A., & Freyd, J. J. (2006). Research participants telling the truth about their lives: The ethics of asking and not asking about abuse. *American Psychologist, 61*(3), 218–226. https://doi.org/10.1037/0003-066X.61.3.218

Benatar, S. R. (2002). Reflections and recommendations on research ethics in developing countries. *Social Science & Medicine, 54*(7), 1131–1141. https://doi.org/10.1016/S0277-9536(01)00327-6.

Bracken-Roche, D., Bell, E., Macdonald, M. E., & Racine, E. (2017). The concept of 'vulnerability' in research ethics: An in-depth analysis of policies and guidelines. *Health Research Policy and Systems, 15*(1), 8. https://doi.org/10.1186/s12961-016-0164-6.

Brotsky, S. R., & Giles, D. (2007). Inside the "pro-ana" community: A covert online participant observation. *Eating Disorders: The Journal of Treatment & Prevention, 15*(2), 93–109. https://doi.org/10.1080/10640260701190600.

Burbules, N. (2009). Privacy and new technologies. In D. M. Mertens & P. E. Ginsberg (Eds.), *The handbook of social research ethics* (pp. 537–549). London: Sage.

Castellano, M. B. (2014). Ethics of aboriginal research. In W. Teary, J. S. Gordon, & A. D. Renteln (Eds.), *Global bioethics and human rights. Contemporary issues* (pp. 273–288). Lanham: Rowman & Littlefield.

Christensen, L. (1988). Deception in psychological research: When is its use justified? *Personality and Social Psychology Bulletin, 14*(4), 664–675.

Cohen, J. J. (1988). *Statistical power analysis for the behavioral sciences* (2nd ed.). Hillsdale: Erlbaum.

Cromer, L. D., Freyd, J. J., & Binder, A. K. (2006). What's the risk in asking? Participant reaction to trauma history questions compared with reaction to other personal questions. *Ethics and Behavior, 16*(4), 347–362. https://doi.org/10.1207/s15327019eb1604_5.

Damianakis, T., & Woodford, M. R. (2012). Qualitative research with small connected communities: Generating new knowledge while upholding research ethics. *Qualitative Health Research.* https://doi.org/10.1016/j.fsigen.2012.10.002.

DePrince, A. P., & Freyd, J. J. (2006). Costs and benefits of being asked about trauma history. *Journal of Trauma Practice, 3*(4), 23–35. https://doi.org/10.1300/J189v03n04_02.

Diener, E., & Crandall, R. (1978). *Ethics in social and behavioral research.* Oxford: University of Chicago Press.

Douglas, H. (2014). Values in Social Science. In N. Cartwright & E. Montuschi (Eds.), *Philosophy of the social sciences. A new introduction* (pp. 162–184). Oxford: Oxford University Press.

Elam, G., & Fenton, K. A. (2003). Researching sensitive issues and ethnicity: Lessons from sexual health. *Ethnicity and Health, 8*(1), 15–27. https://doi.org/10.1080/13557850303557.

Ellis, C. (2014). Telling secrets, revealing lives. Relational ethics in research with intimate others. *Qualitative Inquiry, 19*(1), 3–29. https://doi.org/10.1177/2F1077800406294947.

Eysenbach, G., & Till, J. E. (2001). Ethical issues in qualitative research on internet communities. *British Medical Journal, 323*(7321), 1103–1105. https://doi.org/10.1136/bmj.323.7321.1103.

Friedman Ross, L., Loup, A., Nelson, R. M., Botkin, J. R., Smith, G. R., & Gehler, S. (2009). The challenges of collaboration for academic and community partners in a research partnership: Points to consider. *Journal of Empirical Research on Human Research Ethics, 5*(1), 19–31. https://doi.org/10.1525/2Fjer.2010.5.1.19.

Goffman, A. (2014). *On the run: Fugitive life in an American city.* Chicago: University of Chicago Press.

Goldacre, B. (2009). *Bad science.* London: Fourth Estate.

Guillemin, M., & Gillam, L. (2004). Ethics, reflexivity, and "ethically important moments" in research. *Qualitative Inquiry, 10*(2), 261–280. https://doi.org/10.1177/2F1077800403262360.

Haggerty, D. C. (2004). Ethics creep: Governing social science research in the name of ethics. *Qualitative Sociology, 27*(4), 391–414.

Heinrich, J., Heine, S. J., & Norenzayan, A. (2010). The weirdest people in the world? *Behavioral and Brain Sciences, 33*(2–3), 61–83. https://doi.org/10.1017/S0140525X0999152X.

Hey, J. D. (1998). Experimental economics and deception: A comment. *Journal of Economic Psychology, 19*, 397–401.

Hjortskov, N., Rissén, D., Blangsted, A. K., et al. (2004). The effect of mental stress on heart rate variability and blood pressure during computer work. *European Journal of Applied Physiology, 92*, 84–89. https://doi.org/10.1007/s00421-004-1055-z.

Hughes, J. (ed.) (2010). European Textbook on Ethics in Research. Brussels, European Union Directorate-General for Research.

Illes, J., Kirschen, M. P., Edwards, E., Stanford, L. R., Bandettini, P., Cho, M. K., Ford, P. J., Glover, G. H., Kulynych, J., Macklin, R., Michael, D. B., Wolf, S. M., et al. (2006). Incidental findings in brain imaging research. *Science, 311*, 783–784. https://doi.org/10.1126/science.1124665.

Israel, M. (2015). *Research ethics and integrity for social scientists* (2nd ed.). London: Sage.

Johnson, D., & Bullock, M. (2009). The ethics of data archiving: Issues from four perspectives. In D. Mertens & P. Ginsberg (Eds.), *The handbook of social research ethics* (pp. 214–228). London: Sage.

Khanlou, N., Koh, J. G., & Mill, C. (2008). Cultural identity and experiences of prejudice and discrimination of Afghan and Iranian immigrant youth. *International Journal of Mental Health and Addiction, 6*(4), 494–513. https://doi.org/10.1007/s11469-008-9151-7.

Kimmel, A. J. (1998). In defense of deception. *American Psychologist, 53*, 803–805. https://psycnet.apa.org/doi/10.1037/0003-066X.53.7.803.

Lawson, E. (2001). Informational and relational meanings of deception: Implications for deception methods in research. *Ethics & Behavior, 11*(2), 115–130. https://doi.org/10.1207/S15327019EB1102_1.

Liang, B., & Lu, H. (2006). Conducting fieldwork in China: Observations on collecting primary data regarding crime, law, and the criminal justice system. *Journal of Contemporary Criminal Justice, 22*(2), 157–172. https://doi.org/10.1177/2F1043986206286918.

Lubet, S. (2015). Ethics on the run. *The New Rambler Review*, May, 15–34. https://ssrn.com/abstract=2611742

Mark, M. M., & Lanz-Watson, A. L. (2011). Experiments and quasi-experiments in field settings. In A. T. Panter & S. K. Sterba (Eds.), *Handbook of ethics in quantitative evaluation* (pp. 185–209). New York: Routledge.

Marshall, R. D., Spitzer, R. L., Vaughan, R., Mellman, L. A., MacKinon, R. A., & Roose, S. P. (2001). Assessing the subjective experience of being a participant in psychiatric research. *American Journal of Psychiatry, 158*, 319–321.

McAreavey, R., & Das, C. (2013). A delicate balancing act: Negotiating with gatekeepers for ethical research when researching minority communities. *International Journal of Qualitative Methods, 12*(1), 113–131. https://doi.org/10.1177/2F160940691301200102.

Mertens, D., & Ginsberg, P. E. (Eds.). (2009). *The handbook of social research ethics*. London: Sage.

Mikesell, L., Bromley, E., & Khodyakov, D. (2013). Ethical community-engaged research: A literature review. *American Journal of Public Health, 103*(12), e7–e14. https://doi.org/10.2105/AJPH.2013.301605.

Miller, F. G., Mello, M. M., & Joffe, S. (2008). Incidental findings in human subjects research: What do investigators owe research participants? *Journal of Law, Medicine & Ethics, 36*(2), 271–279. https://doi.org/10.1111/j.1748-720X.2008.00269.x.

Mitcham, C. (2003). Co-responsibility for research integrity. *Science and Engineering Ethics, 9*(2), 273–290. https://doi.org/10.1007/s11948-003-0014-0.

Mustanski, B. (2011). Ethical and regulatory issues with conducting sexuality research with LGBT adolescents: A call to action for a scientifically informed approach. *Archives of Sexual Behavior, 40*(4), 673–686. https://doi.org/10.1007/s10508-011-9745-1.

Newman, E., Walker, E. A., & Gefland, A. (1999). Assessing the ethical costs and benefits of trauma-focused research. *Law, Ethics, and Psychiatry in the General Hospital, 21*(3), 187–196.

Nie, J. B. (2001). Is informed consent not applicable in China? Intellectual flaws of the "cultural difference argument". *Formosan Journal of Medical Humanities, 2*(1–2), 67–74.

Nolen, A. L., & van der Putten, J. (2007). Action research in education: Addressing gaps in ethical principles and practices. *Educational Researcher, 36*(7), 401–407. https://doi.org/10.3102/2F0013189X07309629.

Numan, D., & Di Domenico, M. L. (2012). Market research and the ethics of big data. *International Journal of Market Research, 55*(4), 505–520. https://doi.org/10.2501/2FIJMR-2013-015.

O'Reilly, M., & Parker, N. (2012). 'Unsatisfactory saturation': A critical exploration of the notion of saturated sample sizes in qualitative research. *Qualitative Research, 13*(2), 190–197. https://doi.org/10.1177/2F1468794112446106.

Oliver, P. (2010). *The student's guide to research ethics* (2nd ed.). Berkshire: Open University Press.

Ortmann, A., & Hertwig, R. (2002). The costs of deception: Evidence from psychology. *Experimental Economics, 5*, 111–131. https://doi.org/10.1023/A:1020365204768.

Parker, L., London, A. J., & Aronson, J. D. (2012). Incidental findings in the use of DNA to identify human remains: An ethical assessment. *Forensic Science International Genetics, 7*(2), 221–229. https://doi.org/10.1016/j.fsigen.2012.10.002.

Payne, S. L. (2000). Challenges for research ethics and moral knowledge construction in the applied social sciences. *Journal of Business Ethics, 26*(4), 307–318. https://doi.org/10.1023/A:1006173106143.

Perkman, M., Tartari, V., McKelvey, M., Autio, E., Broström, A., D'Esti, P., Fini, R., Geuna, A., Grimaldi, R., Hughes, A., Krabel, S., Kitson, M., Lierena, P., Lissoni, F., Salder, A., & Sobrero, M. (2013). Academic engagement and commercialization: A review of the literature on university-industry relations. *Research Policy, 42*(2), 423–442. https://doi.org/10.1016/j.respol.2012.09.007.

Ran, M. S., Ziang, M. Z., Chan, C. L. W., Leff, J., Simpson, P., Huang, M. S., Shan, Y. H., & Li, S. G. (2003). Effectiveness of psychoeducational intervention for rural Chinese families experiencing schizophrenia. *Social Psychiatry and Psychiatric Epidemiology, 38*(2), 69–75. https://doi.org/10.1007/s00127-003-0601-z.

Rangle, L., Embry, T., & MacLeon, T. (2001). Active and passive consent: A comparison of actual research with children. *Ethical Human Sciences and Services, 3*(1), 23–31. https://doi.org/10.1891/1523-150X.3.1.23.

Schenk, K. D., & RamaRao, S. (2016). Ethical considerations of conducting research among children and young people affected by HIV: A view from an ethics review board. In P. Liamputtong (Ed.), *Children and young people living with HIV/AIDS: A cross-cultural perspective* (pp. 445–457). Cham: Springer.

Schneider, C. (2015). *The censor's hand: The misregulation of human-subject research.* Cambridge: The MIT Press.

Scott, A. (2017). *My IRB nightmare. Slate Star Codex.* Retrieved March 17, 2020, from http://slatestarcodex.com/2017/08/29/my-irb-nightmare/#comments

Szala-Meneok, K. (2009). Ethical research with older adults. In D. M. Mertens & P. E. Ginsberg (Eds.), *The handbook of social research ethics* (pp. 507–517). London: Sage.

Tangwa, G. B. (2014). Ethics, human rights, and sexual/reproductive health in Africa: Exploratory sociocultural considerations. In W. Teary, J. S. Gordon, & A. D. Renteln (Eds.), *Global bioethics and human rights: Contemporary issues* (pp. 160–173). Lanham: Rowman & Littlefield.

Tolich, M. (2014). What can Milgram and Zimbardo teach ethics committees and qualitative researchers about minimizing harm? *Research Ethics, 10*(2), 86–96. https://doi.org/10.1177/2F1747016114523771.

Van Bommel, M., van Prooijen, J.-W., Elffers, H., & van Lange, P. A. M. (2014). Intervene to be seen: The power of a camera in attenuating the bystander effect. *Social Psychological and Personality Science, 5*(4), 459–466. https://doi.org/10.1177/1948550613507958.

Vargas, L. A., & Montaya, M. E. (2009). Involving minors in research. In D. M. Mertens & P. E. Ginsberg (Eds.), *The handbook of social research ethics* (pp. 489–509). London: Sage.

Vernooij, M. W., Arfan Ikram, M., Tanghe, H. L., Vincent, A. J. P. E., Hofman, A., Krestin, G. P., Niessen, W. J., Breteler, M. M. B., & van der Lugt, A. (2007). Incidental findings on brain MRI in the general population. *The New England Journal of Medicine, 358*(18), 1821–1828. https://doi.org/10.1056/NEJMoa070972.

Williamson, G. R., & Prosser, S. (2002). Action research: Politics, ethics and participation. *Journal of Advanced Nursing, 40*(5), 587–593. https://doi.org/10.1046/j.1365-2648.2002.02416.x.

Author Index

© The Author(s) 2020
J. Bos, *Research Ethics for Students in the Social Sciences*,
https://doi.org/10.1007/978-3-030-48415-6

Subject Index

A

Academic
    capitalism, 177, 178, 191
    freedom, 25, 26, 166, 168, 170, 212
    labor market, 212
Academic Anonymous, 200
Accountability, 18, 26, 39, 40, 45, 58,
        136, 188
Added value, 176, 216, 235
Adjunctification, 212, 217
Agencies, 16, 19, 26, 39, 95, 96, 132, 168, 190
American Anthropological Association
        (AAA), 139
American Psychological Association
        (APA), 59, 195
American Sociological Association
        (ASA), 165
Animal care, 39
Anonymity, 38, 48, 136, 151, 156, 158–159,
        161–163, 166, 168, 170, 236, 252,
        264, 268
Anonymization, 154, 157–162, 167,
        263, 264
Anti-vaccination, 261
Arena, 13, 21
AskAcademia, 87, 130, 133
Asymmetrical treatment, 237
Authorship
    ghost, 95
    gift, 208
Autonomy, 23, 26, 40–43, 48, 153, 155, 157,
        161, 162, 168, 178, 183, 188, 244, 248
Avoiding harm, 231

B

Behavioral Insights Team (BIT), 22
Benefits, 3, 39, 40, 42, 48, 96, 155, 165, 189,
        191, 231–235, 237, 239
Bias
    confirmation, 120–122, 130, 136,
        137, 254
    disciplinary, 185
    editorial, 129–131, 137
    funding, 181, 204, 254
    myside, 120, 136
    publication, 128–131, 136, 137, 254
    reporting, 137, 188
    reviewer, 129–131, 185
    submission, 128
Big data, 157, 251
Bioethics, 195
Blind protocols, 165
Bloom's taxonomy of knowledge-based
        learning, 19
Board of complaint, 41
Bribery, 183
British Royal Society, 10, 11
Bureaucratization, 26

C

Canvassing, 84
Chance findings, 155, 202, 255
Cheating
    crisis, 91, 92, 106
Code of conduct, 39–41, 43, 63, 66
Communism (communalism), 20, 21, 23

© The Author(s) 2020
J. Bos, *Research Ethics for Students in the Social Sciences*,
https://doi.org/10.1007/978-3-030-48415-6

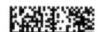